Signals and Communication Technology

T0213967

Shlomo Engelberg

Digital Signal Processing

An Experimental Approach

 Springer

Shlomo Engelberg, Ph.D.
Jerusalem College of Technology
Electronics Department
21 HaVaad HaLeumi Street
Jerusalem
Israel

ISBN 978-1-84996-730-3   e-ISBN 978-1-84800-119-0

DOI 10.1007/978-1-84800-119-0

British Library Cataloguing in Publication Data
Engelberg, Shlomo
    Digital signal processing : an experimental approach. -
    (Signals and communication technology)
    1. Signal processing - Digital techniques
    I. Title
    621.3'822

Cover design: WMX Design GmbH, Heidelberg, Germany

Printed on acid-free paper

9 8 7 6 5 4 3 2 1

springer.com

This work is dedicated to:

My wife, Yvette, and to our children:
Chananel, Nediva, and Oriya.

Preface

The field known as digital signal processing (DSP) has its roots in the 1940s and 1950s, and got started in earnest in the 1960s [10]. As DSP deals with how computers can be used to process signals, it should come as no surprise that the field's growth parallels the growth in the use of the computer. The modern development of the Fast Fourier Transform in 1965 gave the field a great push forward. Since the 1960s, the field has grown by leaps and bounds.

In this book, the reader is introduced to the theory and practice of digital signal processing. Much time is spent acquainting the reader with the mathematics and the insights necessary to master this subject. The mathematics is presented as precisely as possible; the text, however, is meant to be accessible to a third- or fourth-year student in an engineering program.

Several different aspects of the digital signal processing "problem" are considered. Part I deals with the *analysis* of discrete-time signals. First, the effects of sampling and of time-limiting a signal are considered. Next, the spectral analysis of signals is considered. Both the DFT and the FFT are considered, and their properties are developed to the point where the reader will understand both their mathematical content and how they can be used in practice.

After discussing spectral analysis and very briefly considering the spectral analysis of random signals, we move on to Part II. We take a break from the most mathematical parts of DSP, and we consider how one takes an analog signal and converts it into a digital one and how one takes a digital signal and converts it into an analog signal. We present many different types of converters in moderate depth.

After this tour of analog to digital and digital to analog converters, we move on to the third part of the book—and consider the *design* and *analysis* of digital filters. The Z-transform is developed carefully and then the properties, advantages, and disadvantages of infinite impulse response (IIR) and finite impulse response (FIR) filters are explained.

Over the last several years, MATLAB® and Simulink® have become ubiquitous in the engineering world. They are generally good tools to use when one wants to analyze and implement the mathematical techniques of signal processing. They are used throughout this book as tools of analysis and as platforms for designing, implementing, and testing algorithms.

Throughout the book, MATLAB and Simulink are used to allow the reader to *experience* DSP. It is hoped that in this way the beautiful mathematics presented will be seen to be part of a practical engineering discipline.

The Analog Devices ADuC841 is used to introduce the practical microprocessor-oriented parts of digital signal processing. Many chapters contain ADuC841-based laboratories—as well as traditional exercises. The ADuC841 is an easy to use and easy to understand, 8052-based microcontroller (or microconverter®, to use Analog Devices' terminology). It is, in many ways, an ideal processor for student use. It should be easy to "transpose" the ADuC841-based laboratories to other microprocessors

It is assumed that the reader is familiar with Fourier series and transforms and has some knowledge of signals and systems. Some acquaintance with probability theory and the theory of functions of a (single) complex variable will allow the reader to use this text to best advantage.

After reading this book, the reader will be familiar with both the theory and practice of digital signal processing. It is to be hoped that the reader will learn to appreciate the way that the many elegant mathematical results presented form the core of an important engineering discipline.

In preparing this work, I was helped by the many people who read and critically assessed it. In particular, Prof. Aryeh Weiss of Bar Ilan University, and Moshe Shapira and Beni Goldberg of the Jerusalem College of Technology provided many helpful comments. My students at the Jerusalem College of Technology and at Bar Ilan University continue to allow me to provide all my course materials in English, and their comments and criticisms have improved this work.

My family has supported me throughout the period during which I spent too many nights completing this work—and without their support, I would neither have been able to, nor would I have desired to, produce this book. Though many have helped with this undertaking and improved this work, any mistakes that remain are my own.

Shlomo Engelberg
Jerusalem, Israel

Contents

Part II Analog to Digital and Digital to Analog Converters

The Analysis of Discrete-time Signals

1

Understanding Sampling

Summary. In Part I, we consider the analysis of discrete-time signals. In Chapter 1, we consider how discretizing a signal affects the signal's Fourier transform. We derive the Nyquist sampling theorem, and we give conditions under which it is possible to reconstruct a continuous-time signal from its samples.

Keywords. sample-and-hold, Nyquist sampling theorem, Nyquist frequency, aliasing, undersampling.

1.1 The Sample-and-hold Operation

Given a function $g(t)$, if one samples the function when $t = nT_s$ and one holds the sampled value until the next sample comes, then the result of the sampling procedure is the function $\tilde{g}(t)$ defined by

$$\tilde{g}(t) \equiv g(nT_s), \quad nT_s \le t < (n+1)T_s.$$

It is *convenient* to *model* the sample-and-hold operations as two separate operations. The first operation is sampling the signal by multiplying the signal by a train of delta functions

$$\Delta(t) \equiv \sum_{n=-\infty}^{\infty} \delta(t - nT_s).$$

A sampler that samples in this fashion—by multiplying the signal to be sampled by a train of delta functions—is called an *ideal sampler*. The multiplication of $g(t)$ by $\Delta(t)$ leaves us with a train of impulse functions. The areas of the impulse functions are equal to the samples of $g(t)$. After ideal sampling, we are left with

$$\sum_{n=-\infty}^{\infty} g(nT_s)\delta(t - nT_s).$$

The information that we want about the function is here, but the extraneous information—like the values the function takes between sampling times—is gone.

Next, we would like to take this ideally sampled signal and hold the values between samples. As we have a train of impulses with the correct areas, we need a "block" that takes an impulse with area A, transforms it into a rectangular pulse of height A that starts at the time at which the delta function is input to the block, and persists for exactly T_s seconds. A little bit of thought shows that what we need is a linear, time-invariant (LTI) filter whose impulse response, $h(t)$, is 1 between $t = 0$ and $t = T_s$ and is zero elsewhere.

Let us define the Fourier transform of a function, $y(t)$, to be

$$Y(f) = \mathcal{F}(y(t))(f) \equiv \int_{-\infty}^{\infty} e^{-2\pi j f t} y(t) \, dt.$$

It is easy enough to calculate the Fourier transform of $h(t)$—the frequency response of the filter—it is simply

$$H(f) = \frac{1 - e^{-2\pi j T_s f}}{2\pi j f}.$$

(See Exercise 2.)

1.2 The Ideal Sampler in the Frequency Domain

We have seen how the "hold" part of the sample-and-hold operation behaves in the frequency domain. How does the ideal sampler look? To answer this question, we start by considering the Fourier series associated with the function $\Delta(t)$.

1.2.1 Representing the Ideal Sampler Using Complex Exponentials: A Simple Approach

Proceeding formally and not considering what is meant by a delta function too carefully[1], let us consider $\Delta(t)$ to be a periodic function. Then its Fourier series is [7]

$$\Delta(t) = \sum_{n=-\infty}^{\infty} c_n e^{2\pi j n t / T_s},$$

and

$$c_n = \frac{1}{T_s} \int_{-T_s/2}^{T_s/2} e^{-2\pi j n t / T_s} \Delta(t) \, dt = \frac{1}{T_s} \cdot 1 = F_s, \qquad F_s \equiv 1/T_s.$$

---

[1] The reader interested in a careful presentation of this material is referred to [19].

F_{s}, the reciprocal of T_{s}, is the frequency with which the samples are taken. We find that

$$\Delta(t) = F_{\mathrm{s}} \sum_{n=-\infty}^{\infty} e^{2\pi j n F_{\mathrm{s}} t}.$$

1.2.2 Representing the Ideal Sampler Using Complex Exponentials: A More Careful Approach

In this section, we consider the material of Section 1.2.1 in greater detail and in a more rigorous fashion. (This section can be skipped without loss of continuity.) Rather than proceeding formally, let us try to be more careful in our approach to understanding $\Delta(t)$. Let us start by "building" $\Delta(t)$ out of complex exponentials. Consider the sums

$$h_N(t) \equiv \sum_{n=-N}^{N} e^{2\pi j n F_{\mathrm{s}} t}. \tag{1.1}$$

We show that as $N \to \infty$ the function $h_N(t)$ tends, in an interesting sense, to a constant multiple of $\Delta(t)$.

Rewriting (1.1) and making use of the properties of the geometric series, we find that for $t \neq m/F_{\mathrm{s}}$,

$$\begin{aligned}
h_N(t) &\equiv \sum_{n=-N}^{N} e^{2\pi j n F_{\mathrm{s}} t} \\
&= e^{-2\pi j N t} \sum_{n=0}^{2N} e^{2\pi j n F_{\mathrm{s}} t} \\
&= e^{-2\pi j N t} \frac{1 - e^{2\pi j (2N+1) F_{\mathrm{s}} t}}{1 - e^{2\pi j F_{\mathrm{s}} t}} \\
&= \frac{\sin(\pi(2N+1)F_{\mathrm{s}} t)}{\sin(\pi F_{\mathrm{s}} t)}.
\end{aligned}$$

When $t = m/F_{\mathrm{s}}$, it is easy to see that $h_N(t) = 2N + 1$. Considering the limits of $h_N(t)$ as $t \to mT_{\mathrm{s}}$, we find that $h_N(t)$ is a continuous function. (It is not hard to show that $h_N(t)$ is actually an analytic function. See Exercise 6.)

The defining property of the delta function is that when one integrates a delta function times a continuous function, the integration returns the value of the function at the point at which the delta function tends to infinity. Let us consider the integral of $h_N(t)$ times a continuous function $g(t)$. Because $h_N(t)$ is a combination of functions that are periodic with period $T_s \equiv 1/F_{\mathrm{s}}$, so is $h_N(t)$. We consider the behavior of $h_N(t)$ on the interval $[-T_{\mathrm{s}}/2, T_{\mathrm{s}}/2)$. Because of the periodicity of $h_N(t)$, the behavior of $h_N(t)$ on all other such intervals must be essentially the same.

Let us break the integral of interest into three pieces. One piece will consist of the points near $t = 0$—where we know that the sum becomes very large as N becomes very large. The other pieces will consist of the rest of the points. We consider

$$\int_{-T_{\mathrm{s}}/2}^{T_{\mathrm{s}}/2} h_N(t)g(t)\,\mathrm{d}t = \int_{-1/N^{2/5}}^{1/N^{2/5}} h_N(t)g(t)\,\mathrm{d}t + \int_{-T_{\mathrm{s}}/2}^{-1/N^{2/5}} h_N(t)g(t)\,\mathrm{d}t$$
$$+ \int_{1/N^{2/5}}^{T_{\mathrm{s}}/2} h_N(t)g(t)\,\mathrm{d}t.$$

Considering the value of the last integral, we find that

$$\int_{1/N^{2/5}}^{T_{\mathrm{s}}/2} h_N(t)g(t)\,\mathrm{d}t = \int_{1/N^{2/5}}^{T_{\mathrm{s}}/2} \sin(\pi(2N+1)F_{\mathrm{s}}t)(g(t)/\sin(\pi F_{\mathrm{s}}t))\,\mathrm{d}t.$$

We would like to show that this integral tends to zero as $N \to \infty$. Note that if $g(t)$ is nicely behaved in the interval $[1/N^{2/5}, T_{\mathrm{s}}/2]$ then, since $\sin(\pi F_{\mathrm{s}}t)$ is never zero in this interval, $g(t)/\sin(\pi F_{\mathrm{s}}t)$ is also nicely behaved in the interval. Let us consider

$$\lim_{N\to\infty} \int_{1/N^{2/5}}^{T_{\mathrm{s}}/2} \sin(\pi(2N+1)F_{\mathrm{s}}t)r(t)\,\mathrm{d}t$$

where $r(t)$ is assumed to be once continuously differentiable. Making use of integration by parts, we find that

$$\lim_{N\to\infty} \left| \int_{1/N^{2/5}}^{T_{\mathrm{s}}/2} \sin(\pi(2N+1)F_{\mathrm{s}}t)r(t)\,\mathrm{d}t \right|$$

$$= \lim_{N\to\infty} \left| \left(r(t) \frac{-\cos(\pi(2N+1)F_{\mathrm{s}}t)}{\pi(2N+1)F_{\mathrm{s}}} \right|_{1/N^{2/5}}^{T_{\mathrm{s}}/2} \right.$$
$$\left. + \int_{1/N^{2/5}}^{T_{\mathrm{s}}/2} \frac{\cos(\pi(2N+1)F_{\mathrm{s}}t)}{\pi(2N+1)F_{\mathrm{s}}} r'(t)\,\mathrm{d}t \right) \right|$$

$$\leq \lim_{N\to\infty} \left(\frac{2\max_{1/N^{2/5}\leq t\leq T_{\mathrm{s}}/2} |r(t)|}{\pi(2N+1)F_{\mathrm{s}}} \right.$$
$$\left. + \frac{(T_{\mathrm{s}}/2 - 1/N^{2/5})\max_{1/N^{2/5}\leq t\leq T_{\mathrm{s}}/2} |r'(t)|}{\pi(2N+1)F_{\mathrm{s}}} \right).$$

Assuming that for small t we know that $|r(t)| < K_1/|t|$ and $|r'(t)| < K_2/|t|^2$—as is the case for $g(t)/\sin(\pi F_{\mathrm{s}}t)$—we find that as $N \to \infty$, the value of the integral tends to zero. By identical reasoning, we find that as $N \to \infty$,

$$\int_{-T_{\mathrm{s}}/2}^{-1/N^{2/5}} h_N(t)g(t)\,\mathrm{d}t \to 0.$$

Thus, everything hinges on the behavior of the integral

$$\int_{-1/N^{2/5}}^{1/N^{2/5}} h_N(t)g(t)\,\mathrm{d}t.$$

That is, everything hinges on the values of $g(t)$ near $t = 0$.

Let us assume that $g(t)$ is four times continuously differentiable at $t = 0$. Then, we know that $g(t)$ satisfies

$$g(t) = g(0) + g'(0)t + g''(0)t^2/2 + g'''(0)t^3/6 + g^{(4)}(\xi)\xi^4/24$$

for some ξ between 0 and t [17]. This allows us to conclude that

$$\lim_{N\to\infty} \int_{-1/N^{2/5}}^{1/N^{2/5}} \frac{\sin(\pi(2N+1)F_s t)}{\sin(\pi F_s t)} g(t)\,\mathrm{d}t$$

$$= \lim_{N\to\infty} \int_{-1/N^{2/5}}^{1/N^{2/5}} \frac{\sin(\pi(2N+1)F_s t)}{\sin(\pi F_s t)}$$

$$\times \left(g(0) + g'(0)t + g''(0)t^2/2 + g'''(0)t^3/6 + g^{(4)}(\xi)\xi^4/24 \right)\,\mathrm{d}t.$$

We claim that the contribution to the limit from the terms

$$g'(0)t + g''(0)t^2/2 + g'''(0)t^3/6 + g^{(4)}(\xi)$$

is zero. Because the function multiplying $g(t)$ is even, the contribution made by $g'(0)t$ must be zero. The product of the two functions is odd, and the region is symmetric. Similarly, the contribution from $g'''(0)t^3/6$ must be zero.

Next consider

$$\int_{-1/N^{2/5}}^{1/N^{2/5}} \frac{\sin(\pi(2N+1)F_s t)}{\sin(\pi F_s t)} g^{(4)}(\xi)\frac{\xi^4}{24}\,\mathrm{d}t = \int_{-1/N^{2/5}}^{1/N^{2/5}} h_N(t)g^{(4)}(\xi)(\xi^4/24)\,\mathrm{d}t.$$

Clearly $g^{(4)}(\xi)(\xi^4/24)$ is of order $(1/N^{2/5})^4$ for $t \in [-1/N^{2/5}, 1/N^{2/5}]$. Considering (1.1) and making use of the triangle inequality:

$$\left| \sum_{n=-N}^{N} a_k \right| \leq \sum_{n-N}^{N} |a_k|,$$

it is clear that

$$|h_N(t)| \leq \sum_{n=-N}^{N} 1 = 2N + 1.$$

As the interval over which we are integrating is of width $2/N^{2/5}$, it is clear that the contribution of this integral tends to zero as $N \to \infty$. Let us consider

$$\int_{-1/N^{2/5}}^{1/N^{2/5}} \frac{\sin(\pi(2N+1)F_{\mathrm{s}}t)}{\sin(\pi F_{\mathrm{s}}t)} g''(0)t^2/2 \, \mathrm{d}t.$$

It is clear that

$$\left| \int_{-1/N^{2/5}}^{1/N^{2/5}} \frac{\sin(\pi(2N+1)F_{\mathrm{s}}t)}{\sin(\pi F_{\mathrm{s}}t)} g''(0)t^2/2 \, \mathrm{d}t \right| \leq 2(2N+1) \int_{0}^{1/N^{2/5}} |g''(0)|t^2/2 \, \mathrm{d}t$$

$$= 2(2N+1)|g''(0)|(1/N^{2/5})^3/6.$$

As $N \to \infty$, this term also tends to zero. Thus, to calculate the integral of interest, all one needs to calculate is

$$\lim_{N\to\infty} \int_{-1/N^{2/5}}^{1/N^{2/5}} \frac{\sin(\pi(2N+1)F_{\mathrm{s}}t)}{\sin(\pi F_{\mathrm{s}}t)} g(0) \, \mathrm{d}t.$$

Substituting $u = \pi(2N+1)F_{\mathrm{s}}t$, we find that we must calculate

$$\frac{1}{\pi(2N+1)F_{\mathrm{s}}} \int_{-(2N+1)/N^{2/5}}^{(2N+1)/N^{2/5}} \frac{\sin(u)}{\sin(u/(2N+1))} g(0) \, \mathrm{d}u.$$

Note that as $N \to \infty$, we find that $u/(2N+1)$ is always small in the region over which we are integrating. It is, therefore, easy to justify replacing $\sin[u/(2N+1)]$ by $u/(2N+1)$. After making that substitution, we must calculate

$$\lim_{N\to\infty} \frac{1}{\pi(2N+1)F_{\mathrm{s}}} \int_{-(2N+1)/N^{2/5}}^{(2N+1)/N^{2/5}} \frac{\sin(u)}{u/(2N+1)} g(0) \, \mathrm{d}u = \frac{g(0)}{\pi F_{\mathrm{s}}} \int_{-\infty}^{\infty} \frac{\sin(u)}{u} \, \mathrm{d}u.$$

This last integral is well known; its value is π [3, p. 193]. We find that

$$\lim_{N\to\infty} \int_{-T_{\mathrm{s}}/2}^{T_{\mathrm{s}}/2} h_N(t)g(t) \, \mathrm{d}t = T_{\mathrm{s}}g(0).$$

Thus, as $N \to \infty$, the function $h_N(t)$ behaves like $T_{\mathrm{s}}\delta(t)$ in the region $[-T_{\mathrm{s}}/2, T_{\mathrm{s}}/2]$. By periodicity, we find that as $N \to \infty$,

$$h_N(t) \to T_{\mathrm{s}} \sum_{n=-\infty}^{\infty} \delta(t - nT_{\mathrm{s}}).$$

We have found that

$$\Delta(t) = \sum_{n=-\infty}^{\infty} \delta(t - nT_{\mathrm{s}}) = F_{\mathrm{s}} \sum_{n=-\infty}^{\infty} e^{2\pi jnF_{\mathrm{s}}t}.$$

1.2.3 The Action of the Ideal Sampler in the Frequency Domain

The ideal sampler takes a function, $g(t)$, and multiplies it by another "function," $\Delta(t)$. Thus, in the frequency domain it convolves the Fourier transform of $g(t)$, $G(f)$, with the Fourier transform of $\Delta(t)$.

What is the Fourier transform of $\Delta(t)$? Proceeding with impunity, we state that

$$\mathcal{F}(\Delta(t))(f) = F_s \sum \mathcal{F}(e^{2\pi jnF_s t})(f) = F_s \sum_{n=-\infty}^{\infty} \delta(f - nF_s).$$

It is (relatively) easy to see that when one convolves a function with a shifted delta function one "moves" the center of the function to the location of the "center" of the delta function. Thus, the convolution of $G(f)$ with the train of delta functions leaves us with copies of the Fourier transform of $G(f)$ that are spaced every F_s Hz. We find that the Fourier transform of the ideally sampled function is

$$\mathcal{F}(g(t)\Delta(t))(f) = F_s \sum_{n=-\infty}^{\infty} G(f - nF_s). \tag{1.2}$$

Let us assume that $G(f)$ is band-limited:

$$G(f) = 0, \qquad |f| > F.$$

Consider, for example, $G(f)$ as given in Figure 1.1. When considering the sum of shifted versions of $G(f)$, we find that two possibilities exist. If F is sufficiently small, then the different copies of $G(f)$ do not overlap, and we can see each copy clearly. See Figure 1.2. If, on the other hand, F is too large, then there is overlap between the different shifted versions of $G(f)$, and it is no longer possible to "see" $G(f)$ by simply looking at the sum of the shifted version of $G(f)$.

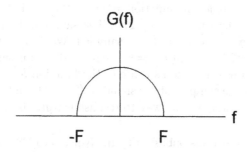

Fig. 1.1. The spectrum of the band-limited function $G(f)$

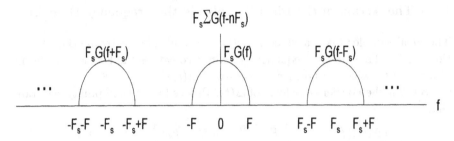

Fig. 1.2. The spectrum of the ideally sampled function when there is no overlap between copies

If the copies of $G(f)$ do not overlap, then by low-pass filtering the signal one can recover the original signal. When will the Fourier transforms not overlap? Considering Figure 1.2, it is clear that in order to prevent overlap, we must require that $F < F_s - F$. That is, we must require that

$$F < F_s/2.$$

That is, we must require that the highest frequency in the signal be less than half of the sampling frequency. This is the content of the celebrated *Nyquist sampling theorem*, and one half the sampling rate is known as the *Nyquist frequency*[2].

1.3 Necessity of the Condition

We have shown that if the highest frequency in a signal is less than half the sampling rate, then it is possible to reconstruct the signal from its samples. It is easy to show that if the highest frequency in a signal is greater than or equal to the half the sampling frequency, then it is not generally possible to reconstruct the signal.

Consider, for example, the function $g(t) = \sin[2\pi F t]$. Let us take $2F$ samples per second at the times $t = k/(2F)$. The sampling frequency is *exactly* twice the frequency of the signal being sampled. We find that the samples of the signal are $g[k/(2F)] = \sin(\pi k) = 0$. That is, all of our samples are zeros. As these samples are the same as those of the function $h(t) = 0$, there is no way to distinguish the samples of the signal $\sin(2\pi F t)$ from those of the signal $h(t) \equiv 0$. There is, therefore, no way to reconstruct $g(t)$ from its samples.

---

[2] The sampling theorem was published by H. Nyquist in 1928, and was proved by C.E. Shannon in 1949. See [18] for more information about the history of the Nyquist sampling theorem.

1.4 An Interesting Example

Suppose that $g(t) = \cos(2\pi F_s t)$ and that one is sampling F_s times per second. As we are violating the Nyquist criterion—we are sampling at the same frequency as the highest frequency present—we should *not* find that the sampled-and-held signal looks similar to the original signal.

Let us use Fourier analysis (which is certainly *not* the easy way here) to see what the output of the sample-and-hold element will be. The Fourier transform of our signal is two delta functions, each of strength $1/2$, located at $\pm F_s$. After sampling, these become a train of delta functions located at nF_s each with strength F_s. After passing this signal through the "hold block" we find that all the delta functions at $nF_s, n \neq 0$ are multiplied by zero and are removed. The delta function at $f = 0$ is multiplied by T_s, and we are left with $F_s T_s \delta(f) = \delta(f)$. This is the transform of $\tilde{g}(t) = 1$. Thus, we find that after the sample-and-hold operation the cosine becomes a "one." See Figure 1.3. (Show that the output of the sample-and-hold element is one in a second way. Consider only the sample-and-hold operation, and do not use Fourier transforms at all.)

1.5 Aliasing

Suppose that one samples a cosine of frequency F at the sampling rate F_s where $F_s > F > F_s/2$ and then "reconstructs" the signal using an ideal low-pass filter that passes all frequencies up to $F_s/2$. What frequency will one see at the output of the filter?

In Figure 1.4, we see the spectrum of the unsampled cosine and of the ideally sampled cosine. If we low-pass filter the sampled cosine using a low-pass filter whose cut-off frequency is $F_s/2$ (and that amplifies by a factor of T_s) then at the output of the filter we will have two impulses of strength $1/2$. They will be located at $F_s - F$ and at $-F_s + F$. This is the Fourier transform of $\cos(2\pi(F_s - F)t)$. We find that the reconstructed signal appears at the wrong frequency. This phenomenon is known as *aliasing*. In order to avoid this problem, one must place an analog low-pass filter whose cut-off frequency is less than or equal to the Nyquist frequency *before* the input to the sampling circuitry. Such a filter is known as an *anti-aliasing filter*.

1.6 The Net Effect

Consider what happens when one has an ideal sampler followed by a hold "circuit" of the type described previously. The ideal sampler makes copies of the spectrum of the signal every F_s Hz. The hold circuit then filters this new signal. How does the filtering work? Let us consider $H(f)$ again:

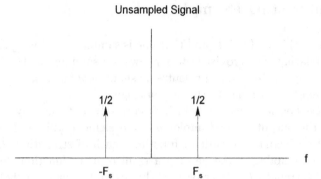

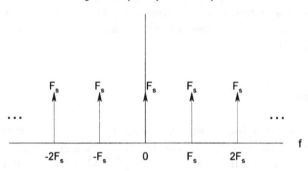

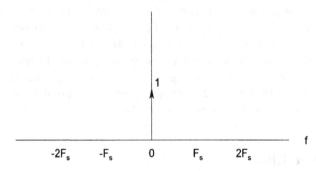

Fig. 1.3. A simple example of aliasing

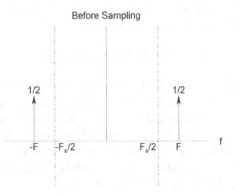

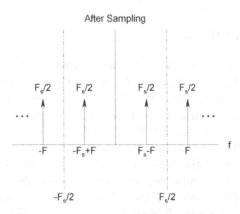

Fig. 1.4. A more general example of aliasing

$$H(f) = \frac{1 - e^{-2\pi j T_s f}}{2\pi j f}.$$

A simple application of the triangle inequality, $|a + b| \leq |a| + |b|$, shows that

$$|H(f)| \leq \frac{1}{\pi |f|}.$$

This is a low-pass filter of sorts.

The spectrum at the output of the sample-and-hold element is

$$V_{\text{out}}(f) = \frac{1 - e^{-2\pi j f T_s}}{2\pi j f} F_s \sum_{-\infty}^{\infty} V_{\text{in}}(f - nF_s)$$

$$= e^{-\pi j f T_s} \frac{\sin(\pi f / F_s)}{\pi (f / F_s)} \sum_{-\infty}^{\infty} V_{\text{in}}(f - nF_s).$$

For relatively small values of f we find that $e^{-\pi j f T_s}$ and $\sin(\pi f/F_s)/(\pi f/F_s)$ are both near 1. When f is small we see that

$$V_{out}(f) \approx V_{in}(f), \quad |f| \ll F_s.$$

Let us consider how the rest of the copies of the spectrum are affected by this filtering. At $f = nF_s$, the sine term is zero. Thus, near multiples of the sampling frequency the contribution of the copies is small. In fact, *as long as the sampling frequency is much greater than the largest frequency in the signal, the contribution that the copies of the spectrum will make to the spectrum of the output of the sample-and-hold element will be small.* If the sampling rate is not high enough, this is not true. See Exercise 7.

1.7 Undersampling

Suppose that one has a real signal all of whose energy is located between the frequencies F_1 and F_2 (and $-F_2$ and $-F_1$) where $F_2 > F_1$. A naive application of the Nyquist sampling theorem would lead one to conclude that in order to preserve the information in the signal, one must sample the signal at a rate exceeding $2F_2$ samples per second. This, however, need not be so.

Consider the following example. Suppose that one has a signal whose energy lies between 2 and 4 kHz (exclusive of the endpoints). If one samples the signal at a rate of 4,000 sample per second, then one finds that the spectrum is copied into non-overlapping regions. Thus, after such sampling it is still possible to recover the signal. Sampling at a frequency that is less than the Nyquist frequency is called *undersampling*. Generally speaking, in order to be able to reconstruct a signal from its samples, one must sample the signal at a frequency that exceeds twice the signal's *bandwidth*.

1.8 The Experiment

1. Write a program for the ADuC841 that causes the microcontroller to sample a signal 1,000 times each second. Use channel 0 of the ADC for the sampling operation.
2. Have the program move the samples from the ADC's registers to the registers that "feed" DAC 0. This will cause the samples to be output by DAC 0.
3. Connect a signal generator to the ADC and an oscilloscope to the DAC.
4. Use a variety of inputs to the ADC. Make sure that some of the inputs are well below the Nyquist frequency, that some are near the Nyquist frequency, and that some exceed the Nyquist frequency. Record the oscilloscope's output.

1.9 The Report

Make sure that your report includes the program you wrote, the plots that you captured, and an explanation of the extent to which your plots agree with the theory described in this chapter.

1.10 Exercises

1. Suppose $g(t) = \sin(2\pi F_s t)$ and one uses a sample-and-hold element that samples at the times

$$t = nT_s, n = 0, 1, \ldots, \qquad F_s = 1/T_s.$$

   Using Fourier transforms, calculate what the sampled-and-held waveform will be.

2. Show that the frequency response of a filter whose impulse response is

$$h(t) = \begin{cases} 1 & 0 \le t < T_s \\ 0 & \text{otherwise} \end{cases}$$

   is

$$H(f) = \begin{cases} \frac{1-e^{-2\pi j f T_s}}{2\pi j f} & f \ne 0 \\ T_s & f = 0 \end{cases}.$$

3. Show that $H(f)$—the frequency response of the "hold element"—can be written as

$$H(f) = \begin{cases} e^{-j\pi T_s f} \frac{\sin(\pi T_s f)}{\pi f} & f \ne 0 \\ T_s & f = 0 \end{cases}.$$

4. Let $H(f)$ be given by the function

$$H(f) = \begin{cases} 1 & 2{,}200 < |f| < 2{,}800 \\ 0 & \text{otherwise} \end{cases}.$$

   If one uses an ideal sampler to sample $h(t)$ every $T_s = 0.5\,\text{ms}$, what will the spectrum of the resulting signal be?

5. Show that the spectrum of an ideally sampled signal as given in (1.2) is periodic in f and has period F_s.

6. Show that the function

$$f(t) = \begin{cases} \frac{\sin(\pi(2N+1)t)}{\sin(\pi t)} & t \ne k \\ 2N+1 & t = k \end{cases}$$

   is
   a) Periodic with period 1.
   b) Continuous on the whole real line.

Note that as both the numerator and the denominator are analytic functions and the quotient is continuous, the quotient must be analytic. (This can be proved using Morera's theorem [3, p. 133], for example.)

7. Construct a Simulink® model that samples a signal 100 times per second and outputs the samples to an oscilloscope. Input a sinewave of frequency 5 Hz and one of frequency 49 Hz. You may use the "zero-order hold" block to perform the sample-and-hold operation. Can you identify the 5 Hz sinewave from its sampled version? What about the 49 Hz sinewave? Explain why the oscilloscope traces look the way they do.

2

Signal Reconstruction

Summary. We have seen that if one samples a signal at more than twice the highest frequency contained in the signal, then it is possible, *in principle*, to reconstruct the signal. In this chapter, we consider the reconstruction problem from a somewhat more practical perspective.

Keywords. reconstruction, Taylor series.

2.1 Reconstruction

From what we described in Chapter 1, it would seem that all that one needs to do to reconstruct a signal is to apply an ideal low-pass filter—a low-pass filter whose frequency response is one up to the Nyquist frequency and zero for all higher frequencies—to the sampled signal.

This is not quite true. Let us consider a *practical* sampling system—a system in which the sample-and-hold element is a single element. In such a system, one does not see perfect copies of the signal in the frequency domain. Such "pure copies" are found only when viewing an ideally sampled signal—a signal multiplied by a train of delta functions. As we saw in Chapter 1, the spectrum at the output of the sample-and-hold element is

$$V_{\text{out}}(f) = e^{-\pi j f T_s} \frac{\sin(\pi f / F_s)}{\pi(f/F_s)} \sum_{n=-\infty}^{\infty} V_{\text{in}}(f - nF_s).$$

The sampling operation creates copies of the spectrum; the hold operation filters *all* the copies somewhat. Even given an ideal low-pass filter that completely removes all the copies of the spectrum centered at $f = nF_s, n \neq 0$, one finds that the *baseband copy* of the spectrum, the copy centered at 0 Hz, is

$$V_{\text{low-pass}}(f) = e^{-\pi j f T_s} \frac{\sin(\pi f / F_s)}{\pi(f/F_s)} V_{\text{in}}(f).$$

As we know [17] that

$$\sin(x)/x \approx 1 - x^2/3, \qquad |x| << 1,$$

we find that if f is not too large, the magnitude of the Fourier transform of the low-passed version of the output of the sample-and-hold element is approximately

$$|V_{\text{low-pass}}(f)| \approx \left(1 - \frac{(\pi f/F_s)^2}{3!}\right)|V_{\text{in}}(f)|.$$

If the sampling frequency is five times greater than the highest frequency present in the signal, then the highest frequency will be attenuated by a factor of approximately

$$\text{attenuation} \approx 1 - (\pi/5)^2/6 = 0.93.$$

That is, even after *ideal filtering*, the output of the sample-and-hold unit may be attenuated by as much as 7%, and the degree of attenuation is *frequency-dependent*.

2.2 The Experiment

1. Build a Simulink® system composed of a sinewave generator, a zero-order hold element (which is equivalent to our sample-and-hold element), a high-order Butterworth filter, and an oscilloscope.
2. Set the zero-order hold to sample 100 times per second, and let the cut-off frequency of the filter be 50 Hz.
3. Input signals that are well below, near, and above the Nyquist frequency. Record the input to, and the output of, the system.

2.3 The Report

In your report, include a description of the Simulink model built, a figure that illustrates the system, and printouts of the oscilloscope input and output for a variety of frequencies.

2.4 Exercises

1. By making use of the fact that the Taylor series that corresponds to $\sin(x)/x$,

$$\frac{\sin(x)}{x} = 1 - x^2/3! + \cdots + (-1)^k x^{2k}/(2k+1)! + \cdots,$$

is an alternating series, prove that

$$\left| \frac{\sin(x)}{x} - (1 - x^2/3!) \right| \leq \frac{1}{120}, \qquad |x| \leq 1.$$

3

Time-limited Functions Are Not Band-limited

Summary. In this chapter, we show that it is impossible for both a function and its Fourier transform to be well localized. We show that if a function is compactly supported—if it is zero outside of some bounded region—then its Fourier transform cannot be compactly supported. Then we show that the "narrower" a function is in the time domain, the more spread out it will be in the frequency domain (and *vice versa*).

Keywords. compact support, analytic function, time-limited, band-limited, uncertainty principle.

3.1 A Condition for Analyticity

The Fourier transform of a function, $g(t)$, is

$$G(f) = \int_{-\infty}^{\infty} e^{-2\pi j f t} g(t) \, dt.$$

Suppose that the function, $g(t)$, is continuous and compactly supported—that $g(t) = 0$ if $|t| \geq T$. Then the Fourier transform of $g(t)$, $G(f)$, is equal to

$$G(f) = \int_{-T}^{T} e^{-2\pi j f t} g(t) \, dt = \int_{-T}^{T} \sum_{k=0}^{\infty} \frac{(-2\pi j f t)^k}{k!} g(t) \, dt.$$

We would like to interchange the order of integration and summation, so we must show that the series converges uniformly. To this end, we consider the absolute value of the terms

$$\frac{(-2\pi j f t)^k}{k!} g(t)$$

when $|t| \leq T$. We find that

$$\left| \frac{(-2\pi jft)^k}{k!} g(t) \right| \leq \frac{(2\pi|f|T)^k}{k!} \max_{|t| \leq T} |g(t)|.$$

We now consider the remainder of the series appearing in the calculation of $G(f)$. We consider the sum

$$\sum_{k=N}^{\infty} \frac{(-2\pi jft)^k}{k!} g(t).$$

We find that

$$\left| \sum_{k=N}^{\infty} \frac{(-2\pi jft)^k}{k!} g(t) \right| \leq \sum_{k=N}^{\infty} \left| \frac{(-2\pi jft)^k}{k!} g(t) \right| \leq \sum_{k=N}^{\infty} \frac{(2\pi|f|T)^k}{k!} \max_{|t| \leq T} |g(t)|.$$

Considering the sum

$$\sum_{k=N}^{\infty} C\frac{x^k}{k!}, \qquad x > 0,$$

we find that

$$\sum_{k=N}^{\infty} C\frac{x^k}{k!} = C\frac{x^N}{(N-1)!} \sum_{k=N}^{\infty} \frac{x^{k-N}}{k(k-1)\cdots N}$$

$$\leq C\frac{x^N}{(N-1)!} \sum_{k=N}^{\infty} \frac{x^{k-N}}{(k-N)!}$$

$$= \frac{x^N}{(N-1)!} e^x.$$

As $N \to \infty$ this sum tends to zero. Applying this result to our sum, we find that

$$\left| \sum_{k=N}^{\infty} \frac{(-2\pi jft)^k}{k!} g(t) \right| \leq \max_{|t| \leq T} |g(t)| \frac{(2\pi|f|T)^N}{(N-1)!} e^{2\pi|f|T}.$$

As $N \to \infty$ this sum tends to zero uniformly in t, and the sum that appears in the calculation of $G(f)$ is uniformly convergent. This allows us to interchange the order of summation and integration in the integral that defines $G(f)$. Interchanging the order of summation and integration, we find that

$$G(f) = \sum_{k=0}^{\infty} f^k \frac{(-2\pi j)^k}{k!} \int_{-T}^{T} t^k g(t) \, dt.$$

As $G(f)$ is represented by a convergent Taylor series, it is analytic [3, p. 159]. That is, the Fourier transform of a compactly supported function is analytic. (In order to show that the value of the remainder of the series tends to zero uniformly, we made use of the fact that $g(t)$ is zero for $|t| > T$. When $g(t)$ is not time-limited, it *is* possible for $G(f)$ to be band-limited.)

3.2 Analyticity Implies Lack of Compact Support

It is well known and easy to prove that a non-constant analytic function cannot be constant along any smooth curve [3, p. 284]. Thus, as long as $G(f)$ is not identically zero, it cannot be zero on any interval. Thus, a non-zero time-limited function—a non-zero function that is compactly supported in time—cannot be band-limited. The same basic proof shows that band-limited functions cannot be time-limited. (See Exercise 1.)

3.3 The Uncertainty Principle

We now prove that a function cannot be well localized in time and frequency. We showed above that if a function is completely localized in time, it cannot be completely localized in frequency. We now extend this result.

We prove that

$$\left(\int_{-\infty}^{\infty} t^2 |f(t)|^2 \, dt / E \right) \left(\int_{-\infty}^{\infty} f^2 |F(f)|^2 \, df / E \right) \geq 1/(16\pi^2) \qquad (3.1)$$

where

$$E \equiv \int_{-\infty}^{\infty} |f(t)|^2 \, dt \stackrel{\text{Parseval}}{=} \int_{-\infty}^{\infty} |F(f)|^2 \, df.$$

The normalized integrals in (3.1) measure the degree of localization of a function and its Fourier transform. (See Exercise 3.) The bigger either number is, the less localized the relevant function is.

To prove (3.1), we consider the integral

$$\int_{-\infty}^{\infty} t f(t) \frac{df(t)}{dt} \, dt.$$

The Cauchy-Schwarz inequality for integrals [18] shows that

$$\left| \int_{-\infty}^{\infty} t f(t) \frac{df(t)}{dt} \, dt \right| \leq \sqrt{\int_{-\infty}^{\infty} t^2 |f(t)|^2 \, dt} \sqrt{\int_{-\infty}^{\infty} \left| \frac{df(t)}{dt} \right|^2 \, dt}. \qquad (3.2)$$

Let us evaluate the leftmost integral. Making the (relatively mild) assumption that

$$\lim_{|t| \to \infty} \sqrt{|t|} f(t) = 0,$$

we find that

$$\int_{-\infty}^{\infty} t f(t) \frac{df(t)}{dt} \, dt = \int_{-\infty}^{\infty} t \frac{df^2(t)/2}{dt} \, dt$$

$$= t f^2(t)/2 \Big|_{-\infty}^{\infty} - \int_{-\infty}^{\infty} f^2(t)/2 \, dt$$

$$= - \int_{-\infty}^{\infty} f^2(t)/2 \, dt.$$

Let us now evaluate the other integral of interest in (3.2). Making use of Parseval's theorem [7] and the fact that the Fourier transform of the derivative of a function is $2\pi j f$ times the Fourier transform of the original function, we find that

$$\int_{-\infty}^{\infty} \left| \frac{\mathrm{d}f(t)}{\mathrm{d}t} \right|^2 \mathrm{d}t = (2\pi)^2 \int_{-\infty}^{\infty} f^2 |F(f)|^2 \, \mathrm{d}f.$$

Squaring both sides of (3.2) and combining all of our results, we find that

$$\left(\int_{-\infty}^{\infty} t^2 |f(t)|^2 \, \mathrm{d}t / E \right) \left(\int_{-\infty}^{\infty} f^2 |F(f)|^2 \, \mathrm{d}f / E \right) \geq 1/(16\pi^2).$$

3.4 An Example

Let us consider the function $g(t) = \mathrm{e}^{-|t|}$. It is well known [7] that

$$G(f) = \frac{2}{(2\pi f)^2 + 1}.$$

In this case,

$$E = \int_{-\infty}^{\infty} \left(\mathrm{e}^{-|t|} \right)^2 \mathrm{d}t = \int_{-\infty}^{\infty} \mathrm{e}^{-2|t|} \, \mathrm{d}t = 1.$$

By making use of integration by parts twice, it is easy to show that

$$\int_{-\infty}^{\infty} t^2 \mathrm{e}^{-2|t|} \, \mathrm{d}t = \frac{1}{2}.$$

What remains is to calculate the integral

$$\int_{-\infty}^{\infty} \frac{4f^2}{[(2\pi f)^2 + 1]^2} \, \mathrm{d}f.$$

One way to calculate this integral is to make use of the method of residues. (For another method, see Exercise 6.)

Let C_R be the boundary of the upper half-disk of radius R traversed in counter-clockwise direction. That is, we start the curve from $-R$, continue along the real axis to $+R$, and then leave the real axis and traverse the upper semicircle in the counter-clockwise direction. Because the order of the numerator is two greater than that of the denominator, it is easy to show that as $R \to \infty$, the contribution of the semicircle tends to zero.

We find that

$$\int_{-\infty}^{\infty} \frac{4f^2}{[(2\pi f)^2 + 1]^2} \, \mathrm{d}f = 4 \lim_{R \to \infty} \oint_{C_R} \frac{z^2}{[(2\pi z)^2 + 1]^2} \, \mathrm{d}z.$$

As long as $R > 1$, we find that inside the curve C_R, the integrand has one pole of multiplicity 2 at the point $z = -j/(2\pi)$. Rewriting the integrand as

$$\frac{z^2}{[(2\pi z)^2 + 1]^2} = \frac{1}{(2\pi)^4} \frac{z^2}{[z + j/(2\pi)]^2 [z - j/(2\pi)]^2},$$

it is clear that the residue at $j/(2\pi)$ is

$$\frac{\mathrm{d}}{\mathrm{d}z} \frac{1}{(2\pi)^4} \frac{z^2}{[z + j/(2\pi)]^2}\bigg|_{z=j/(2\pi)} = \frac{-j\pi}{2} \frac{1}{(2\pi)^4}.$$

We find that for all $R > 1$,

$$4 \oint_{C_R} \frac{z^2}{[(2\pi z)^2 + 1]^2} \, \mathrm{d}z = 4 \times 2\pi j \times -\frac{j\pi}{2} \frac{1}{(2\pi)^4} = \frac{1}{4\pi^2}.$$

In particular, we conclude that

$$\int_{-\infty}^{\infty} \frac{4f^2}{[(2\pi f)^2 + 1]^2} \, \mathrm{d}f = \frac{1}{4\pi^2}.$$

All in all, we find that

$$\left(\int_{-\infty}^{\infty} t^2 |f(t)|^2 \, \mathrm{d}t / E \right) \left(\int_{-\infty}^{\infty} f^2 |F(f)|^2 \, \mathrm{d}f / E \right) = \frac{1}{8\pi^2} > \frac{1}{16\pi^2}.$$

This is in perfect agreement with the theory we have developed.

3.5 The Best Function

Which functions achieve the lower bound in (3.1)? In our proof, it is the Cauchy-Schwarz inequality that leads us to the conclusion that the product is greater than or equal to $1/(16\pi^2)$. In the proof of the Cauchy-Schwarz inequality (see, for example, [7]), it is shown that *equality* holds if the two functions whose squares appear in the inequality are constant multiples of one another. In our case, this means that we must find the functions that satisfy

$$\frac{\mathrm{d}f(t)}{\mathrm{d}t} = ct f(t).$$

It is easy to see (see Exercise 2) that the only functions that satisfy this differential equation and that are square integrable[1] are the functions

$$f(t) = De^{ct^2/2}, \qquad c < 0.$$

---

[1] Square integrable functions are functions that satisfy $\int_{-\infty}^{\infty} |f(t)|^2 \, \mathrm{d}t < \infty$.

3.6 Exercises

1. Prove that a non-constant band-limited function cannot be time-limited.
2. Find the solutions of the equation

$$\frac{df}{dt} = ctf(t).$$

   Show that the only solutions of this equation that are square integrable, that is, the only solutions, $f(t)$, for which

$$\int_{-\infty}^{\infty} f^2(t)\, dt < \infty,$$

   are the functions

$$f(t) = De^{ct^2/2}, \qquad c < 0.$$

3. Consider the functions

$$f_W(t) = \frac{1}{\sqrt{W}} \Pi_W(t)$$

   where the function $\Pi_W(t)$ is defined as

$$\Pi_W(t) \equiv \begin{cases} 1 & |t| \leq W/2 \\ 0 & \text{otherwise} \end{cases}.$$

   Show that

$$\mathrm{loc}(W) \equiv \left(\int_{-\infty}^{\infty} t^2 |f_W(t)|^2\, dt / E \right)$$

   is a monotonically increasing function of W. Explain how this relates to the fact that $\mathrm{loc}(W)$ is a measure of the extent to which the functions $f_W(t)$ are localized.
4. Let

$$g(t) = \frac{1}{\sqrt{2\pi}} e^{-t^2/2}.$$

   Calculate $G(f)$, and show that for this Gaussian function the two sides of Inequality (3.1) are actually equal. (One may use a table of integrals to evaluate the integrals that arise. Alternatively, one can make use of the properties of the Gaussian PDF for this purpose.)
5. Explain why, if one makes use of the criterion of this chapter to determine how localized a function is, it is reasonable to say that the function

$$f(t) = \begin{cases} \frac{\sin(t)}{t} & t \neq 0 \\ 1 & t = 0 \end{cases}$$

   is "totally unlocalized."
6. Let $g(t) = e^{-|t|}$ as it is in Section 3.4.

a) Show that for this $g(t)$,

$$\int_{-\infty}^{\infty} f^2 |G(f)|^2 \, \mathrm{d}f = \int_{-\infty}^{\infty} f^2 G(f)^2 \, \mathrm{d}f.$$

b) Calculate

$$\int_{-\infty}^{\infty} f^2 G(f)^2 \, \mathrm{d}f$$

by making use of Parseval's theorem and the properties of the Fourier transform. (You may ignore any "small" problems connected to differentiating $g(t)$.)

4

Fourier Analysis and the Discrete Fourier Transform

Summary. Having discussed sampling and time-limiting and how they affect a signal's spectrum, we move on to the estimation of the spectrum of a signal from the signal's samples. In this chapter, we introduce the discrete Fourier transform, we discuss its significance, and we derive its properties. Then we discuss the family of algorithms know as fast Fourier transforms. We explain their significance, describe how one uses them, and discuss zero-padding and the fast convolution algorithm.

Keywords. Fourier transform, discrete Fourier transform, fast Fourier transform, zero-padding, fast convolution.

4.1 An Introduction to the Discrete Fourier Transform

Often, we need to determine the *spectral content* of a signal—how much of a signal's power is located at a given frequency. In general, this means that we would like to determine the Fourier transform of the signal. Because we are *actually* measuring the signal, we cannot possibly know its values from $t = -\infty$ to $t = +\infty$. We can only know the signal's value in some finite interval. As we generally use a microprocessor to make measurements, even in that interval we only know the signal's value at discrete times—generally at the times $t = nT_s$ where T_s is the sampling period. From this set of time-limited samples, we would like to *estimate* the Fourier transform of the signal.

If all that one knows of a signal is its value in an interval it is clearly impossible to determine the signal's Fourier transform. Something must be assumed about the signal at the times for which no measurements exist. A fairly standard assumption, and the most reasonable in many ways, is that the function is zero outside the region in which it is measured and is reasonably smooth inside this region. We will, once again, consider the consequences of time-limiting the function—of "lopping off the tails" of the function—in Chapter 5.

Recalling that the Fourier transform of a function, $y(t)$, is

$$Y(f) = \mathcal{F}(y(t))(f) \equiv \int_{-\infty}^{\infty} e^{-2\pi jft} y(t)\, dt,$$

we find that given a function, $y(t)$, which is zero outside the region $t \in [0, T]$, we can express its Fourier transform as

$$Y(f) = \int_{0}^{T} e^{-2\pi jft} y(t)\, dt.$$

Suppose that we have only N samples of the function taken at $t = k(T/N)$, $k = 0, \ldots, N - 1$. Then we can estimate the integral by

$$Y(f) \approx \sum_{k=0}^{N-1} e^{-2\pi jfkT/N} y(kT/N)(T/N).$$

If we specialize the frequencies we are interested in to $f = m/T$, $n = 0, \ldots N - 1$, then we find that

$$Y(m/T) \approx (T/N) \sum_{k=0}^{N-1} e^{-2\pi jmk/N} y_k, \qquad y_k = y(kT/N). \qquad (4.1)$$

The discrete Fourier transform (DFT) of the sequence y_k is defined as

$$Y_m = \mathrm{DFT}(\{y_k\})(m) \equiv \sum_{k=0}^{N-1} e^{-2\pi jmk/N} y_k. \qquad (4.2)$$

The value of the DFT is (up to the constant of proportionality T/N) an approximation of the value of the Fourier transform at the frequency m/T.

4.2 Two Sample Calculations

Consider the sequence $\{y_k\}$ given by

$$\{-1, 1, -1, 1\}.$$

As $N = 4$, we find that

$$\begin{aligned}
Y_0 &= e^0(-1) + e^0(1) + e^0(-1) + e^0(1) = 0 \\
Y_1 &= (e^{-\pi j/2})^0(-1) + e^{-\pi j/2}(1) + (e^{-\pi j/2})^2(-1) + (e^{-\pi j/2})^3(1) \\
&= 1(-1) + (-j)(1) + (-1)(-1) + j(1) = 0 \\
Y_2 &= (-1)^0(-1) + (-1)(1) + (-1)^2(-1) + (-1)^3(1) = -4 \\
Y_3 &= j^0(-1) + j^1(1) + j^2(-1) + j^3(1) = 0.
\end{aligned}$$

Now consider the sequence $\{z_k\}$ given by

$$\{1,1,0,0\}.$$

We find that

$$Z_m = \sum_{k=0}^{3} e^{-2\pi jkm/4} y_k = \sum_{k=0}^{3} \left(e^{-j\pi/2}\right)^{km} y_k = \sum_{k=0}^{3} (-j)^{km} y_k.$$

Thus, we find that

$$Z_0 = 2$$

$$Z_1 = \sum_{k=0}^{3} (-j)^k y_k = 1 - j$$

$$Z_2 = \sum_{k=0}^{3} (-j)^{2k} y_k = \sum_{k=0}^{3} (-1)^k y_k = 0$$

$$Z_3 = \sum_{k=0}^{3} (-j)^{3k} y_k = \sum_{k=0}^{3} j^k y_k = 1 + j.$$

4.3 Some Properties of the DFT

We have seen that the DFT is *an* approximation to the Fourier transform. Why use this particular approximation? The short answer is that the DFT has many nice properties, and many of the DFT's properties "mimic" those of the Fourier transform.

The DFT, $\{Y_k\}$, is an N-term sequence that is derived from another N-term sequence $\{y_k\}$. We will shortly show that the mapping from sequence to sequence is invertible. Moreover, the mapping is almost an isometry—almost norm-preserving. Let the l_2 norm of an N-term sequence $\{c_0, \ldots c_{N-1}\}$ be defined as

$$\sqrt{\sum_{k=0}^{N-1} |c_k|^2}.$$

Then the mapping preserves the l_2 norm (except for multiplication by a constant) as well.

To show that the mapping is invertible, it is sufficient to produce an inverse mapping. Consider the value of the sum

$$a_k = \sum_{m=0}^{N-1} e^{2\pi jkm/N} Y_m.$$

We find that

$$a_k = \sum_{m=0}^{N-1} e^{2\pi jkm/N} Y_m$$

$$= \sum_{m=0}^{N-1} e^{2\pi jkm/N} \sum_{n=0}^{N-1} e^{-2\pi jmn/N} y_n$$

$$= \sum_{m=0}^{N-1} \sum_{n=0}^{N-1} e^{-2\pi jmn/N} e^{2\pi jkm/N} y_n$$

$$= \sum_{n=0}^{N-1} y_n \sum_{m=0}^{N-1} e^{-2\pi jm(n-k)/N}.$$

As the inner sum is simply a geometric series, we can sum the series. We find that for $0 \le n \le N-1$,

$$\sum_{m=0}^{N-1} e^{-2\pi jm(n-k)/N} = \begin{cases} N, & n = k \\ \frac{1-e^{2\pi j(n-k)}}{1-e^{-2\pi j(n-k)/N}} = 0, & \text{otherwise} \end{cases} = N\delta_{nk}$$

where $\delta_{nk} = 1$, $n = k$, and $\delta_{nk} = 0$, $n \ne k$. (The function δ_{nk} is known as the Kronecker[1] delta function.) We find that $a_k = Ny_k$. That is, we find that

$$y_k = a_k/N = \frac{1}{N} \sum_{m=0}^{N-1} e^{+2\pi jkm/N} Y_m.$$

This mapping of sequences is known as the *inverse discrete Fourier transform* (IDFT), and it is almost identical to the DFT.

In Section 4.2, we found that the DFT of the sequence $\{-1, 1, -1, 1\}$ is

$$\{Y_m\} = \{0, 0, -4, 0\}.$$

Using the IDFT, we find that

$$y_0 = \frac{1}{4}(-4) = -1$$

$$y_1 = \frac{1}{4} \sum_{m=0}^{3} \left(e^{\pi j/2}\right)^m Y_m = \frac{1}{4}(-1) \cdot (-4) = 1$$

$$y_2 = \frac{1}{4} \sum_{m=0}^{3} \left(e^{\pi j}\right)^m Y_m = \frac{1}{4} 1 \cdot (-4) = -1$$

$$y_3 = \frac{1}{4} \sum_{m=0}^{3} \left(e^{3\pi j/2}\right)^m Y_m = \frac{1}{4}(-1) \cdot (-4) = 1.$$

These are indeed the samples we started with in Section 4.2.

---

[1] Named after Leopold Kronecker (1823–1891) [18].

Let

$$\mathbf{Y} = \begin{bmatrix} Y_0 \\ \vdots \\ Y_{N-1} \end{bmatrix} \text{ and } \mathbf{y} = \begin{bmatrix} y_0 \\ \vdots \\ y_{N-1} \end{bmatrix}$$

where the elements of $\mathbf{Y}$ are the terms in the DFT of the sequence $\{y_0, \ldots, y_{N-1}\}$. Consider the square of the norm of the vector, $\|\mathbf{Y}\|^2$. We find that

$$\|\mathbf{Y}\|^2 \equiv \sum_{m=0}^{N-1} |Y_m|^2$$

$$= \sum_{m=0}^{N-1} Y_m \overline{Y}_m$$

$$= \sum_{m=0}^{N-1} \sum_{k=0}^{N-1} e^{2\pi jkm/N} y_k \sum_{l=0}^{N-1} e^{-2\pi jlm/N} \overline{y}_l$$

$$= \sum_{k=0}^{N-1} \sum_{l=0}^{N-1} y_k \overline{y}_l \sum_{m=0}^{N-1} e^{-2\pi j(k-l)m/N}$$

$$= \sum_{k=0}^{N-1} \sum_{l=0}^{N-1} y_k \overline{y}_l N \delta_{kl}$$

$$= N \sum_{k=0}^{N-1} y_k \overline{y}_k$$

$$= N \sum_{k=0}^{N-1} |y_k|^2$$

$$= N\|\mathbf{y}\|^2.$$

We find that the mapping is almost an isometry. The mapping preserves the norm up to a constant factor. Considering the transform pair $\{y_k\} \leftrightarrow \{Y_m\}$ of Section 4.2, we find that $\|\mathbf{y}\|^2 = 4$ and $\|\mathbf{Y}\|^2 = 16 = 4 \cdot 4$. This is in perfect agreement with the theory we have developed.

Assuming that the y_k are real, we find that

$$Y_{N-m} = \sum_{k=0}^{N-1} e^{-2\pi jk(N-m)/N} y_k$$

$$= \sum_{k=0}^{N-1} e^{-2\pi jk(-m)/N} y_k$$

$$= \sum_{k=0}^{N-1} e^{2\pi jkm/N} y_k$$

$$= \overline{\sum_{k=0}^{N-1} e^{-2\pi jkm/N} y_k}$$

$$= \overline{Y_m}.$$

This shows that $|Y_{N-m}| = |Y_m|$. It also shows that when the y_k are real, the values of Y_m for $m > N/2$ do not contain any new information.

Consider the definition of the DFT:

$$Y_m = \sum_{k=0}^{N} e^{-2\pi jkm/N} y_k.$$

Suppose that we allow m to be any integer—rather than restricting m to lie between 0 and $N-1$. Then we find that

$$Y_{N+m} = \sum_{k=0}^{N} e^{-2\pi jk(N+m)/N} y_k$$

$$= \sum_{k=0}^{N} e^{-2\pi jkN/N} y_k e^{-2\pi jkm/N} y_k$$

$$= \sum_{k=0}^{N} e^{-2\pi jkm/N} y_k.$$

That is, the "extended DFT," $\{\ldots, Y_{-1}, Y_0, Y_1, \ldots\}$, is periodic with period N.

Finally, let us consider the DFT of the circular convolution (or cyclic convolution) of two N-periodic sequences, a_k and b_k. Let the circular convolution of the two sequence be defined as

$$y_k = a_k * b_k \equiv \sum_{n=0}^{N-1} a_n b_{k-n}.$$

(As is customary, we denote the convolution operation by an asterisk.) We find that the DFT of y_k is

$$Y_m = \sum_{k=0}^{N-1} e^{-2\pi jmk/N} y_k$$

$$= \sum_{k=0}^{N-1} e^{-2\pi jmk/N} \sum_{n=0}^{N-1} a_n b_{k-n}$$

$$= \sum_{n=0}^{N-1} a_n \sum_{k=0}^{N-1} e^{-2\pi jmk/N} b_{k-n}$$

$$= \sum_{n=0}^{N-1} e^{-2\pi jmn/N} a_n \sum_{k=0}^{N-1} e^{-2\pi jm(k-n)/N} b_{k-n}.$$

As the sequence $e^{-2\pi jm(k-n)/N}b_{k-n}$ is periodic of period N in the variable k, the second sum above is simply the DFT of the sequence b_k. The first sum is clearly the DFT of a_k. Thus, we have shown that

$$Y_m = A_m B_m. \tag{4.3}$$

We have shown that the DFT of the circular convolution of two periodic sequences is the product of the DFTs of the two sequences.

4.4 The Fast Fourier Transform

The DFT can be used to approximate the Fourier transform. A practical problem with using the DFT is that calculating the DFT of a vector with N elements seems to require $N \times N$ complex multiplications and $(N - 1) \times N$ complex additions. Calculating the DFT seems to require approximately $2N^2$ arithmetical operations.

In 1965, J.W. Cooley and J.W. Tukey published an important work [5] in which they explained how under some conditions one could calculate an N-term DFT by performing on the order of $N \log(N)$ calculations[2]. As the logarithm of N grows much slower than N, this made (and continues to make) the calculation of the DFT of large sequences possible. The algorithms related to this idea are known as fast Fourier transforms (FFTs).

To see how these algorithms work, consider a sequence with N terms, y_k, $k = 0, \ldots N - 1$, and let N be an even number. The sequence's DFT is given by

$$Y_m = \sum_{k=0}^{N-1} e^{-2\pi jkm/N} y_k$$

$$= \sum_{k=0}^{(N/2)-1} e^{-2\pi j(2k)m/N} y_{2k} + \sum_{k=0}^{(N/2)-1} e^{-2\pi j(2k+1)m/N} y_{2k+1}$$

$$= \sum_{k=0}^{(N/2)-1} e^{-2\pi jkm/(N/2)} y_{2k} + e^{-2\pi jm/N} \sum_{k=0}^{(N/2)-1} e^{-2\pi jkm/(N/2)} y_{2k+1}.$$

We have calculated the DFT of y_k by *breaking up* the N-term DFT into two $N/2$-term DFTs. As each DFT takes approximately $2N^2$ terms, each of the smaller DFTs requires approximately $2N^2/4$ operations. Since we have two of the smaller DFTs to calculate, our total operation count is approximately

---

[2] The technique had been used in the past by such scientists as C.F. Gauss [18]. Until the advent of the computer, the technique was not very important. Cooley and Tukey rediscovered the technique and appreciated its importance for digital signal processing.

N^2. Additionally, the final reconstruction step requires N additions and N multiplications; it requires $2N$ arithmetical operations.

If we calculate the DFTs of the $N/2$-length sequences by splitting the sequences, we will have four $N/4$-length DFTs to calculate. That will require approximately $4 \cdot 2(N/4)^2 = N^2/2$ calculations. In the reconstruction phase, we need $2 \cdot 2N/2 = 2N$ additions and multiplications to reconstruct the two $N/2$-length DFTs. We find that if we split the DFT twice, we need approximately $2N^2/2^2 + 2(2N)$ operations to calculate the original N sample DFT. In general, we find that if we split the dataset k times, we need approximately $2N^2/2^k + k2N$ operations to calculate the original DFT. Supposing that $N = 2^M$, then if we perform M splits—the most we can perform—we need approximately $2N^2/2^M + M2N = 2N + 2\log_2(N)N$ operations in order to calculate the DFT of the original N-term sequence. As N is small relative to $N\log(N)$, it is sufficient to say that FFT algorithm requires on the order of $N\log(N)$ operations in order to calculate the DFT of an N-element sequence.

As the IDFT of a sequence is essentially the same as the DFT of the sequence, it is not surprising that there is also a fast inverse DFT known as the inverse fast Fourier transform (IFFT). The IFFT also requires on the order of $N\log(N)$ operations to calculate the IDFT of an N-term sequence. The FFT algorithm described above works properly only if the number of elements in the sequence is a power of two. Generally speaking, FFT algorithms have a requirement that the number of elements must be of some specific form. (In our case, the number of elements must be a power of two—but there are other algorithms, and they have different conditions.)

4.5 A Hand Calculation

Let us calculate the DFT of the sequence

$$\{x_n\} = \{-1, 1, 1, 1, -1, 0, 1, 0\}$$

using the method of the previous section. We subdivide this large calculation into two smaller calculations. We find that we must calculate the DFTs of the sequences

$$\{-1, 1, -1, 1\} \text{ and } \{1, 1, 0, 0\}.$$

The DFT of the first sequence is, as we have seen, $\{Y_m\} = \{0, 0, -4, 0\}$. The DFT of the second sequence is $\{Z_m\} = \{2, 1 - j, 0, 1 + j\}$. According to our rule, the final DFT must be

$$X_m = Y_m + e^{-2\pi jm/N} Z_m.$$

(In order to calculate values of Y_m and Z_m for $m > 3$, we make use of the periodicity of the DFT and of the fact that all the samples are real numbers.) We find that

$$X_0 = Y_0 + Z_0 = 2$$

$$X_1 = Y_1 + e^{-2\pi j/8} Z_1 = \frac{1-j}{\sqrt{2}}(1-j) = -j\sqrt{2}$$

$$X_2 = Y_2 + e^{-2\pi j2/8} Z_2 = -4$$

$$X_3 = Y_3 + e^{-2\pi j3/8} Z_3 = \frac{-1-j}{\sqrt{2}}(1+j) = -j\sqrt{2}$$

$$X_4 = Y_4 + e^{-2\pi j4/8} Z_4 = Y_0 + e^{-2\pi j4/8} Z_0 = -2$$

$$X_5 = X_{8-3} = \overline{X_3} = j\sqrt{2}$$

$$X_6 = \overline{X_2} = -4$$

$$X_7 = \overline{X_1} = j\sqrt{2}.$$

4.6 Fast Convolution

In addition to the fact that the FFT makes it possible to calculate DFTs very quickly, the FFT also enables us to calculate circular convolutions quickly. Let y_n be the circular convolution of two N-periodic sequences, a_n and b_n. Then,

$$y_n = a_n * b_n = \sum_{k=0}^{N-1} a_k b_{n-k}, \qquad n = 0, \ldots, N-1.$$

In principle, each of the N elements of y_n requires N multiplications and $N-1$ additions. Thus, calculating the circular convolution *seems* to require on the order of N^2 calculations. As we saw previously (on p. 34), the DFT of the circular convolution of two sequences is the product of the DFTs of the sequences. Thus, it is possible to calculate the circular convolution by calculating the DFTs of the sequences, calculating the product of the DFTs, and then calculating the IDFT of the resulting sequence. That is, we make use of the fact that

$$a_n * b_n = \text{IDFT}(\{\text{DFT}(\{a_n\})\} \overset{\text{element-wise}}{\times} \text{DFT}(\{b_n\})\})$$

to calculate the circular convolution of two sequences. When performed in this way by an FFT algorithm, the calculation requires on the order of $N\log(N)$ calculations, and not on the order of N^2 calculations.

4.7 MATLAB, the DFT, and You

MATLAB® has a command, `fft`, that calculates the DFT. The command makes use of an FFT algorithm whenever it can. Otherwise, it performs a "brute force" DFT. The `fft` command makes it very easy to calculate a DFT. The code segment of Figure 4.1 compares the DFT of the signal $y(t) = e^{-|t|}$

with its actual Fourier transform, $Y(f) = \frac{2}{(2\pi f)^2+2}$. The output of the program is given in Figure 4.2. (Note that—as predicted—the DFT is symmetric about its center.)

```
% This program demonstrates the use of MATLAB's FFT command.
t = [-1000:1000]/10;           % The array t has the times to be
                               % examined.

y = exp(-abs(t));

subplot(2,2,1)                 % Subplot is used to break the plot
                               % into a 2x2 set of plots.
plot(t,y,'k')
title('The Signal')
z = fft(y);                    % The fft command will perform a
                               % DFT as efficiently as it can.
f = [0:2000] / (2001/10);      % The array f has the frequencies
                               % that were estimated.
subplot(2,2,2)
plot(f,(200.1/2001)*abs(z),'--k')% We use the absolute value of the
                               % DFT.  We are not really
                               % interested in the phase here.
                               % 200.1/2001 is T/N.
title('The DFT')
subplot(2,2,3)
plot(f, 2 ./((2 * pi * f).^2 + 1),'.-k')
title('The Fourier Transform')
subplot(2,2,4)
plot(f, (200.1/2001)*abs(z),'--k',f, 2./((2*pi*f).^2+1),'.-k')
title('A Comparison of the Fourier Transform and the DFT')
print -djpeg comp.jpg
```

Fig. 4.1. The MATLAB program

A few general comments about the MATLAB commands used should make the code easier to understand.

- The command [-1000:1000] causes MATLAB to produce an array whose elements are the integers from −1,000 to 1,000. The division by 10 in the command [-1000:1000]/10 is performed on an element by element basis. Thus, the final array consists of the numbers from −100 to 100, the elements being incremented by 0.1.
- In MATLAB code, a semicolon suppresses printing. (Generally speaking, MATLAB prints the output of each command it is given.)
- The command **abs** calculates the absolute value of each element of the input to the command.

- The command subplot(m,n,k) causes MATLAB to create or refer to a figure with $m \times n$ subfigures. The next subfigure to be accessed will be subfigure k.
- The command plot(x,y) is used to produce plots. When used as shown above—in its (almost) simplest form—the command plots the values of y against those of x.
- In the command plot(x,y,'--k'), the string '--k' tells MATLAB to use a black dashed line when producing the plot. (The string can be omitted entirely. When no control string is present, MATLAB uses its own defaults to choose line styles and line colors.)
- The command title adds a title to the current plot.

MATLAB has several types of help commands, and all of them can be used to learn more about these commands.

A few comments about the MATLAB code, and how it relates to what we have seen about DFTs, are also in order. First of all, note that in the program the value of the DFT returned by MATLAB is multiplied by 200.1/2,001. This is just T/N— the constant of proportionality from (4.1). ($T = 200.1$ because 2,001 samples are taken and the time between samples is 0.1.)

The vector f is defined as f = [0:2000] / (2001/10);. That the vector must contain 2,001 elements is clear. The data fed to the DFT had 2,001 elements, and the length of the DFT is the same as that of the input data. The actual frequencies to which the elements of the DFT correspond are m/T, $m = 0,\ldots,2,000$. As $T = 2,001/10$, the division by 2,001/10 is explained.

4.8 Zero-padding and Calculating the Convolution

Suppose that one would like to calculate the (ordinary) convolution of two non-periodic sequences. That is, given $\{a_0,\ldots,a_{N-1}\}$ and $\{b_0,\ldots,b_{N-1}\}$, one would like to calculate

$$c_k = a_k * b_k \equiv \sum_{n=0}^{N-1} a_n b_{k-n}, \qquad 0 \le k \le 2N - 2$$

where the value of an element of $\{a_k\}$ or $\{b_k\}$ whose index exceeds $N - 1$ or is less than 0 is taken to be zero.

One way to calculate this convolution is to calculate the given sum for $k = 0,\ldots,2N - 2$. This takes on the order of N^2 operations. We have already seen that for circular convolutions, this is not the best that one can do; here, too, one can do better.

Suppose that we almost double the length of each sequence by adding $N - 1$ zeros to the end of the sequence. That is, let us define the sequences by taking

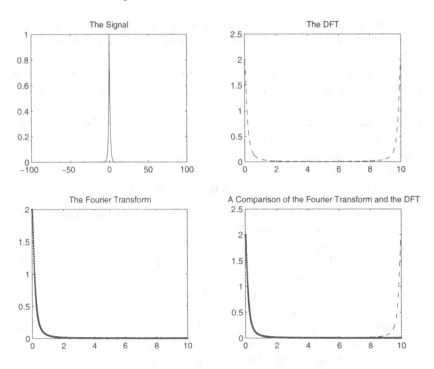

Fig. 4.2. The function, its DFT, its Fourier transform, and a comparison

$$\tilde{a}_k = \begin{cases} a_k & 0 \le k \le N - 1 \\ 0 & N \le k \le 2N - 2 \end{cases}$$

and

$$\tilde{b}_k = \begin{cases} b_k & 0 \le k \le N - 1 \\ 0 & N \le k \le 2N - 2 \end{cases}.$$

We find that the convolution of these two sequences, which we call d_k, is

$$d_k = \tilde{a}_k * \tilde{b}_k = \sum_{n=0}^{2N-2} \tilde{a}_n \tilde{b}_{k-n}.$$

It is easy to see that for $0 \le k \le 2N - 2$, we have $d_k = c_k$.

Now let us consider the *circular convolution* of the two sequences. We find that when the number of zeros added is one less than the length of the original vector, then the circular convolution is exactly the same as the ordinary convolution. If one adds still more zeros to the two original vectors, this will cause additional zeros to trail the desired solution. When calculating the *ordinary* convolution of two sequences, one often chooses to zero-pad the sequences to more than twice their original length to ensure that the number of elements in the final sequence is a power of two. In this way, the convolution can be calculated using the very efficient FFT and IFFT algorithms. This allows us to calculate the convolution in on the order of $N\log(N)$ operations, rather than on the order of N^2 operations.

Consider, for example, the convolution of the two ten-element sequences

$$\{1,\ldots,1\},\text{ and }\{1,2,\ldots,10\}.$$

One way to perform this calculation is to extend both sequences to 32-element sequences where the last 22 elements are all zero. One might give the following commands to MATLAB® to define the sequences

```
a = [ones(size(1:10)) zeros(size(1:22))]
b = [1:10 zeros(size(1:22))]
```

The first command produces a sequence with ten ones and 22 zeros. The second produces a sequence composed of the numbers one through ten followed by 22 zeros. To calculate the convolution, one can give the command

```
ifft(fft(a) .* fft(b))
```

The .* command tells MATLAB to perform an *element by element* multiplication on two sequences. The rest of the command causes MATLAB to calculate the IFFT of the element by element product of the FFTs of the sequences a and b.

4.9 Other Perspectives on Zero-padding

We have seen one reason to zero-pad—it can allow one to calculate an (ordinary) convolution using a circular convolution. There are two more reasons one might want to zero-pad.

Suppose that one has samples, but the number of samples is not a power of two (or some other number of samples that makes using an FFT a possibility). Then one might choose to add samples in order to benefit from the efficiency of the FFT.

Suppose that one adds zeros to a sequence of measurements that one has made. When calculating the DFT of the "inflated" sequence one finds that the maximum frequency of interest is still $F_s/2$—is still the Nyquist frequency. What is changed is the number of frequency bands into which the frequency band $[0, F_s/2]$ is broken. When one zero-pads, one adds more points to the DFT. Thus, *adding zeros in the time domain—zero-padding—leads to interpolation in the spectral domain.*

Sometimes, one adds zeros in order to perform this spectral interpolation. This is a third reason to zero-pad. Note that one has not gained information by zero-padding—one is still calculating essentially the same approximation to the Fourier transform. One is approximating the Fourier transform at more frequency points using the same time-limited, sampled signal.

4.10 MATLAB and the Serial Port

In order to use MATLAB to process data, one must transfer the data from the measurement instrument—in our case, the ADuC841—to MATLAB. We will do this by using the computer's serial port—its UART. Using a serial port is very much like reading from, or writing to, a file. To set up and use a UART from MATLAB, one uses the commands `serial`, `set`, `fopen`, `fprintf`, `fread`, `fclose`, `freeserial`, and `delete`. Use the MATLAB `help` command (or the help window) to get more detail about how these commands are used.

4.11 The Experiment

In this laboratory, we build a system that measures the FFT of a signal more or less on the fly. Use the ADuC841 to sample a signal, and then transmit the measured values to MATLAB across a serial port. Have MATLAB take the measured signal values, perform an FFT on them, and display the magnitude of the FFT for the user.

In greater detail, write a program that causes the ADuC841 to sample 2,000 times per second. Set up the ADuC841's UART to transmit and receive at 57,600 baud. Program the ADuC841 to transmit 512 measurements (1,024 bytes) each time the UART receives the letter "s."

Set up MATLAB to control the computer's serial port. Have MATLAB open the serial port, write the letter "s" to it, and then receive the 1,024 bytes. Once the measurements have been read in, reconstruct the signal measurements and have MATLAB calculate and display the absolute value of the FFT.

4.12 Exercises

1. Consider the following sequences
   - $S_1 \equiv \{1, 0, 0, 0, 0\}$,
   - $S_2 \equiv \{1, -1, 1, -1\}$.
   a) Calculate the DFT of the sequences.
   b) Calculate the norms of the sequences and the transformed sequences.
   c) Compare the norms of the sequences and the transformed sequences, and see if the results agree with the theory presented in the chapter.
2. Why would it be inappropriate to speak of the FFT of the sequence S_1 of Exercise 1?
3. Using the method of Sections 4.4 and 4.5 and the results of Exercise 1, calculate the DFT of

$$\{1, 1, 0, -1, 0, 1, 0, -1\}.$$

4. Note: you may wish to make use of [18] to gather the information necessary to do this exercise.

   a) Give a brief description of the prime-factor algorithm (PFA).

   b) Historically, which came first—the Cooley-Tukey FFT or the PFA?

5. Using MATLAB, calculate the DFT of the sequence

$$\sin(2\pi f n T_\mathrm{s}), \qquad n = 0, \dots, 255$$

   for $T_\mathrm{s} = 1\,\mathrm{ms}$ and $f = 1, 10, 100$, and $500\,\mathrm{Hz}$. Plot the absolute value of the FFT against frequency. (Make sure that the frequency axis in the plot is defined in a reasonable way.) Explain what each plot "says" about the signal.

6. Using MATLAB, calculate the DFT of the sequences $e^{-knT_\mathrm{s}}, n = 0, \dots 255$ for $T_\mathrm{s} = 10\,\mathrm{ms}$ and $k = 1, 2, 4$, and 8. Plot the absolute value of the FFT against frequency. (Make sure that the frequency axis in the plot is defined in a reasonable way.)

7. Calculate the circular convolution of the sequence

$$a_n = \begin{cases} 0, \, n = 0, \dots, 50 \\ 1, \, n = 51, \dots, 204 \\ 0, \, n = 205, \dots, 255 \end{cases}$$

   with itself by using the FFT and the IFFT. Plot the resulting samples against their sample numbers. Submit the MATLAB code along with the answer. (You may wish to make use of the MATLAB commands **ones** and **zeros** to help build up the initial vector.) Note that because of rounding errors, the inverse discrete Fourier transform may have a very small imaginary component. This can lead the plotting routine supplied by MATLAB to give odd results. You may wish to plot the value of the real part of the IDFT calculated by MATLAB. This can be achieved by using the **real** command provided by MATLAB.

8. Calculate the convolution of the two sequences

$$S_1 \equiv \{1, 2, \dots, k, \dots, 20\}$$
$$S_2 \equiv \{1, 4, \dots, k^2, \dots, 400\}$$

   by zero-padding and using the MATLAB commands **FFT** and **IFFT**.

9. Sample a $10\,\mathrm{Hz}$ sinewave at the times $t = kT_\mathrm{s}, k = 0, \dots, 255$ where the sampling period is $10\,\mathrm{ms}$.

   a) Calculate the FFT of this vector of 256 samples. Plot the absolute value of the FFT against frequency. Make sure that the frequency axis is properly normalized.

   b) Now, zero-pad the vector out to a length of 2^{12} samples. Once again, plot the absolute value of the FFT against frequency. Make sure that the frequency axis is properly normalized.

   c) Compare the results of the previous two sections, and explain the similarities and differences that you note.

5

Windowing

Summary. The DFT is an approximation to the Fourier transform of a signal. The input to the DFT is always a sequence of samples taken over a finite time period. This "truncation" of the signal changes the Fourier transform—the spectrum—of the signal. In this chapter, we learn about *windowing*; we learn how to control the effect that truncation has on a signal's spectrum.

Keywords. spectral leakage, sidelobes, windowing, Bartlett, Hann.

5.1 The Problems

By the nature of the way that we make measurements, we are always looking at a finite-length sample of any signal. Let us define the function

$$\Pi_T(t) \equiv \begin{cases} 1 & |t| < T/2 \\ 0 & |t| \geq T/2 \end{cases}.$$

It is easy to show [7] that the Fourier transform of $\Pi_T(t)$ is

$$\mathcal{F}(\Pi_T(t))(f) = \begin{cases} T & f = 0 \\ T\frac{\sin(\pi f T)}{\pi f T} & f \neq 0 \end{cases} \equiv T\mathrm{sinc}(fT). \tag{5.1}$$

When considering a T second snippet of a signal $s(t)$, one is essentially considering $\Pi_T(t)s(t)$. From the standard properties of the Fourier transform, we find that

$$\mathcal{F}(\Pi_T(t)s(t))(f) = S(f) * [T\mathrm{sinc}(fT)]$$

where the asterisk is once again used to represent the convolution operator.

Consider what the result of this operation is when the signal is a cosine—when $s(t) = \cos(2\pi F t)$. The Fourier transform of the signal is

$$S(f) = (1/2)[\delta(f + F) + \delta(f - F)].$$

As convolution with a delta function copies the function being convolved to the location of the delta function, we find that

$$\mathcal{F}(\Pi_T(t)s(t))(f) = (T/2)[\text{sinc}((f + F)T) + \text{sinc}((f - F)T)].$$

That is, the nice, clean delta functions are "smeared out" into sinc functions (and this is in perfect accord with the theory of Chapter 3). See Figure 5.1 for an example of such smearing when $T = 0.2\,\text{s}$ and $F = 20\,\text{Hz}$. Note that the width of the central lobe of the sinc function is $10 = 2/0.2$. That is, the width of the central lobe is twice the reciprocal of the amount of time over which the measurement is performed. In general, $width \times T \approx 2$. (This is a kind of "uncertainty principle." See Section 3.3 for a more precisely stated uncertainty principle.)

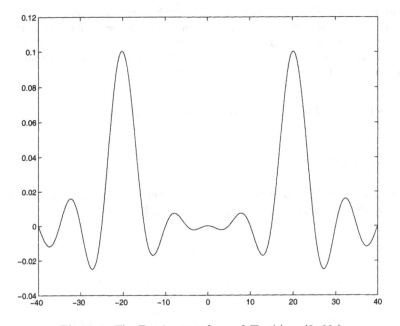

Fig. 5.1. The Fourier transform of $\Pi_{0.2}(t)\cos(2\pi 20t)$

There is a second difficulty. Time limiting not only smears the Fourier transform around the "correct" frequency, but also causes the Fourier transform to extend across all frequencies. In addition to the central lobe, there are now additional lobes, *sidelobes*, as well. The smearing of the Fourier transform that is caused by time limiting a signal is commonly referred to as *spectral leakage*.

We would like to cure the local smearing *and* the appearance of energy at frequencies other than the frequency we are trying to measure. Unfortunately, we *cannot* take care of both problems at once. Generally speaking, what helps with one problem exacerbates the other. Let us see why.

5.2 The Solutions

The Fourier transform of $\Pi_T(t)$ decays relatively slowly—like $1/f$. Why? Because $\Pi_T(t)$ is not continuous. Discontinuous functions—functions with jumps—have a lot of energy at high frequencies. When one multiplies a signal—even a very smooth signal—by $\Pi_T(t)$, one generally forces the function to be discontinuous at the endpoints, $t = \pm T/2$. This discontinuity is reflected in the signal's Fourier transform by its slow decay. To take care of this problem, one must see to it that the sampled values approach zero continuously at the edges of the interval in which the measurement is being made.

Suppose that the function that one has measured is $\Pi_T(t)s(t)$. How can one smooth the edges of this function? What one must do is multiply by a smooth function, $w(t)$, whose support is located in the region $t \in [-T/2, T/2]$ and for which $w(T/2) = w(-T/2) = 0$. Multiplying $\Pi_T(t)s(t)$ by such a function should get rid of some of the high-frequency energy that we saw previously. Of course a function that goes to zero smoothly inside this region will tend to emphasize the region near $t = 0$ more than it was emphasized previously. One can consider this emphasis to be something that shrinks the effective time over which the measurement was taken. As we have seen (in Section 3.3) that the shorter the time period over which the measurement is taken, the more spread out the Fourier transform of the resulting signal will be, we must expect that after multiplication by $w(t)$, the Fourier transform of the measured signal will have less energy at high frequencies—the sidelobes will be smaller, but the Fourier transform will have a more smeared central lobe.

The functions $w(t)$ are known as *window functions*, and the process of multiplying the sampled function by a window function is known as *windowing* the data. There are several standard window functions. Each function strikes a somewhat different balance between the width of the central lobe and the rate at which the high-frequency component decays.

5.3 Some Standard Window Functions

We now present three standard window functions; there are *many* more. When characterizing window functions, we consider:

1. The width of the central peak of the window function's Fourier transform. The width is defined as the distance between the first zero to the left of the central peak and the first zero to the right of the central peak.

2. The rate at which the sidelobes of the Fourier transform of the window function decay.

5.3.1 The Rectangular Window

The window of the previous section, $w(t) = \Pi_T(t)$, is called the rectangular window. Its Fourier transform is

$$\mathcal{F}(\Pi_T(t))(f) = T\mathrm{sinc}(fT).$$

The first zero to the left of the central maximum occurs at $f = -1/T$, and the first zero to the right of the central maximum occurs at $f = 1/T$. Thus, the width of the central lobe is $2/T$. Additionally, as

$$|T\mathrm{sinc}(fT)| = \left|\frac{\sin(\pi fT)}{\pi f}\right| \leq \frac{1}{\pi|f|},$$

we find that the sidelobes decay like $1/f$. Thus, the rectangular window has a narrow main lobe, but its high-frequency terms decay slowly.

5.3.2 The Triangular Window

The triangle function is defined as

$$\Lambda_T(t) = \begin{cases} (t + T/2)/(T/2), & -T/2 \leq t \leq 0 \\ (T/2 - t)/(T/2), & 0 < t \leq T/2 \\ 0, & |t| > T/2 \end{cases} = \frac{2}{T}\Pi_{T/2}(t) * \Pi_{T/2}(t).$$

As the Fourier transform of a convolution is the product of the Fourier transforms of the functions, by making use of (5.1) we find that

$$\mathcal{F}(\Lambda_T(t))(f) = \frac{2}{T}\left(\frac{T}{2}\right)^2 \mathrm{sinc}^2(fT/2) = \frac{T}{2}\mathrm{sinc}^2(fT/2).$$

As the first zeros of $\mathrm{sinc}(fT/2)$ are at $f = \pm 2/T$, the transform has a main lobe that is twice as wide as the "unenhanced" window we used previously. However, the high-frequency terms decay like $1/f^2$ now.

The window based on this idea is known as the *Bartlett window*.

5.4 The Raised Cosine Window

Consider the function

$$w(t) = \Pi_T(t)(1 + \cos(2\pi t/T)).$$

At $t = \pm T/2$, this function is not only continuous; it is once continuously differentiable.

It is not hard to show (see Exercise 3) that the Fourier transform of this function is

$$T\frac{\text{sinc}(fT)}{1 - (fT)^2}.$$

Note that there are zero-canceling poles at $f = -1/T, 0$, and $f = 1/T$. Thus, the zeros that bound the main lobe are located at $f = \pm 2/T$, and the width of the main lobe is $4/T$—just as it is for the Bartlett window. However, the decay at high frequencies is $1/f^3$. Once again, we can trade a wide main lobe for rapid decay of high frequencies.

The window based on this idea is known as the Hann or Hanning window. (But note that MATLAB® uses the names Hann and Hanning to define two slightly different windows.) There are many other window functions. We will not discuss them.

5.5 A Remark on Widths

We have defined the width of the main lobe to be the distance between the zeros of the main lobe. There are other ways of defining the width of the main lobe—and some of the other definitions distinguish between lobes more effectively than our simple definition.

5.6 Applying a Window Function

On a conceptual level, it is easiest to understand how windowing samples affects the spectral content of the sampled signal by considering multiplying the continuous-time signal by a continuous-time window. In practice, one generally samples the input signals "as is." After sampling the signals, one multiplies the samples by the relevant values of the window function. That is, rather than sampling $x(t)w(t)$, one samples $x(t)$, and then multiplies the sampled values $x(nT_s)$ by the samples of the window function, $w(nT_s)$.

MATLAB has predefined window functions like hann and bartlett. These functions take as their argument the number of samples to be windowed. Suppose that the MATLAB variable x has 256 unwindowed samples of a signal. If one would like to apply a raised cosine (Hann) window to the samples and then calculate the FFT of the resulting sequence, one can give MATLAB the command

```
y = fft(x .* hann(256)');
```

The "prime" (apostrophe) after the function call hann(256) takes the window "vector" and transposes it. This changes the window vector from a column vector into a row vector. (Generally speaking, x is a row vector, and MATLAB requires that both items in an element by element operation have the same dimensions. Also note that the ' operator actually calculates the complex conjugate transpose of a matrix. If one has a complex vector whose unconjugated transpose is needed, one uses the . ' command.)

Suppose that one would like to zero-pad and window a set of samples. Which operation should be carried out first? The *reason* one windows a set of samples is to make sure that there are no "artificial" discontinuities at the ends of the dataset. It is, therefore, clear that one must window a dataset *before* zero-padding. Otherwise, the added zeros will add the artificial jumps that windowing is supposed to prevent.

5.7 A Simple Comparison

Using MATLAB, we generated the input to, and the output from, a DFT when the input was $\cos(2\pi 9.9t)$, $T_s = 0.04$, and 200 samples were taken. We calculated the DFT of the data using a rectangular window of length 200, a Bartlett window of length 200, and a Hann window of length 200. The output of the DFT is given in Figure 5.2. Note that the roll-off—the decay of the high frequencies—is slowest in the rectangular window and fastest in the Hann window.

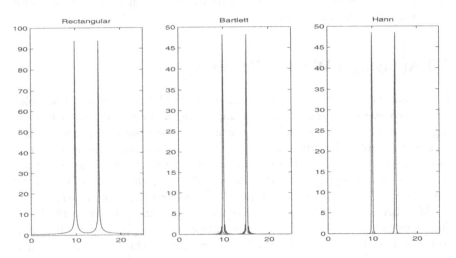

Fig. 5.2. The data after processing by the DFT

5.8 MATLAB's Window Visualization Tool

One nice way to "experience" the different windows is to use the MATLAB window design and analysis tool. Open a MATLAB window, and type in `wintool`. From that point on, everything is self-explanatory.

5.9 The Experiment

Modify the programs written for the experiment of Chapter 4 to use several different window functions. It will probably be simplest to use the built-in window functions provided by MATLAB. Look up the commands `bartlett` and `hann`.

Write programs to use the Bartlett window, the Hann window, and at least one window that has not been discussed in the text. Examine the output of each program when the input to the system is a sinewave. Explain the results in terms of the properties of the window function and the DFT.

5.10 Exercises

1. Use the definition of the Fourier transform to calculate the Fourier transform of $\Pi_T(t)$.
2. Use the MATLAB `wintool` command to compare the Hann window and the Hamming window.
3. Calculate the Fourier transform of the function

$$w(t) = \Pi_T(t)(1 + \cos(2\pi t/T)).$$

4. Calculate the windowed DFT of the sequence $\sin(2\pi f n T_s), n = 0, \ldots, 255$ when $f = 10$ and $T_s = 1\,\text{ms}$, and plot the absolute value of the windowed DFT. Make use of:
   a) the rectangular window,
   b) the Bartlett window, and
   c) the Hanning window.
   Compare the resulting DFTs.
5. Calculate the DFT of the sequence

$$y_k = \sin(2\pi 50 k T_s) + 0.01 \sin(2\pi 60 k T_s), \qquad k = 0, \ldots, 1023, T_s = 1\,\text{ms}.$$

   Plot the absolute value of the DFT (with a properly scaled frequency axis). Then recalculate the DFT using a 1,024-point Hanning window, and plot the absolute value of the resulting sequence (with a properly scaled frequency axis). Explain the differences in the two plots.
6. Explain what *apodization* is and how it relates to windowing.

6

Signal Generation with the Help of MATLAB

Summary. One tool that an engineer often needs is a signal generator. Given a microprocessor with a stable time-base and a nice size block of FLASH memory, it is easy to build a simple signal generator. In this very practically oriented chapter, we show how.

Keywords. signal generator, MATLAB, microprocessor.

6.1 Introduction

The ADuC841 has 64 kB of FLASH[1] program memory from which the microprocessor can read while a program is executing. Our signal generator will be implemented by storing the waveform we would like to output in the FLASH memory and then using one of the ADuC841's timers to control the speed at which the waveform is played back.

Of course, before we can play back a waveform, we must "record" it. We will make all of our "recordings" using MATLAB®.

6.2 A Simple Sinewave Generator

Let us consider how we can "record" a sinewave in a way that will be useful to us. It is our intention to play back our sinewave from the microprocessor's memory.

MATLAB can be used to produce a vector with one full period of a sinewave that has been sampled 256 times. The 256 samples are a good start. Next, the samples must be put in a form that is appropriate for the microprocessor. To do this, the samples must be output in hexadecimal, and they must

---

[1] FLASH memory is non-volatile memory that can be electrically erased and reprogrammed.

correspond to positive voltages (as the ADuC841's DAC converts to positive voltages). Note that MATLAB has a command, dec2hex, that takes an integer as input and produces a string whose characters are the hex digits that correspond to the decimal number that was input. What we would ideally like is to convert the vector of numbers into a set of instructions to the compiler that will make the microprocessor store the values in a meaningful way.

It turns out that getting everything properly formatted is not hard. MATLAB has a command called fprintf that is very similar to the C command of the same name. By using the MATLAB file commands and the fprintf command, one can cause MATLAB to output a file of the form

```
sine:    DB      000H
         DB      00FH
                  .
                  .
                  .
```

This file can then be "pasted" into a program written in the ADuC841's assembly language.

6.3 A Simple White Noise Generator

Not only can one use MATLAB to help write a sinewave generator, one can use it to make a white noise generator. MATLAB has several commands for producing white noise. The one that is most useful for our purposes is the rand command. The rand(size([1:N])) command produces a vector that is N elements long for which each element is (more or less) independent of each other element, and each element is a sample of a value from a random variable that is uniformly distributed over $[0, 1]$. Using fprintf, it is easy to cause MATLAB to generate a long list of (reasonably) uncorrelated numbers.

6.4 The Experiment

1. Cause MATLAB to generate 256 samples of a sinewave.
2. Have MATLAB store the samples in a format that is appropriate for an assembly language program.
3. Write an assembly language program that reads the values of the sinewave and outputs these values to DAC0.
4. Write the program in such a way that the UART can be used to change the frequency of the sinewave. (Have the program change the value(s) of the timer reload register(s) according to the value of the input from the UART.)
5. Examine the output of DAC0, and make sure that the period that is seen corresponds to the predicted period.

6. Next, produce a set of 4,096 samples of random noise and incorporate these values into a program that is similar to the program designed above.

7. Look at the spectrum of the signal produced in this way on an oscilloscope operating in FFT mode. What do you see?

6.5 Exercises

1. Define the term DCO. Please reference a source for the definition.

2. In what way does the MATLAB command `fprintf` differ from C's `fprintf` command?

7

The Spectral Analysis of Random Signals

Summary. When one calculates the DFT of a sequence of measurements of a random signal, one finds that the values of the elements of the DFT do not tend to "settle down" no matter how long a sequence one measures. In this chapter, we present a brief overview of the difficulties inherent in analyzing the spectra of random signals, and we give a quick survey of a solution to the problem—the method of averaged periodograms.

Keywords. random signals, method of averaged periodograms, power spectral density, spectral estimation.

7.1 The Problem

Suppose that one has N samples of a random signal[1], $X_k, k = 0, \ldots, N-1$, and suppose that the samples are independent and identically distributed (IID). Additionally, assume that the random signal is zero-mean—that $E(X_k) = 0$. The expected value of an element of the DFT of the sequence, a_m, is

$$E(a_m) = E\left(\sum_{k=0}^{N-1} e^{-2\pi jkm/N} X_k\right) = 0.$$

Because the signal is zero-mean, so are all of its Fourier coefficients. (All this really means is that the phases of the a_m are random, and the statistical average of such a_m is zero.)

On the other hand, the power at a given frequency is (up to a constant of proportionality) $|a_m|^2$. The expected value of the power at a given frequency

---

[1] In this chapter, capital letters represent random variables, and lowercase letters represent elements of the DFT of a random variable. As usual, the index k is used for samples and the index m for the elements of the DFT. In order to minimize confusion, we do *not* use the same letter for the elements of the sequence and for the elements of its DFT.

is $E(|a_m|^2)$ and is non-negative. If one measures the value of $|a_m|^2$ for some set of measurements, one is measuring the value of a random variable whose expected value is equal to the item of interest. One would expect that the larger N was, the more certainly one would be able to say that the measured value of $|a_m|^2$ is near the theoretical expected value. One would be mistaken.

To see why, consider a_0. We know that

$$a_0 = X_0 + \cdots + X_{N-1}.$$

Assuming that the X_k are real, we find that

$$|a_0|^2 = \sum_{n=0}^{N-1}\sum_{k=0}^{N-1} X_n X_k = \sum_{n=0}^{N-1} X_k^2 + \sum_{n=0}^{N-1}\sum_{k=0}^{N-1,k\neq n} X_n X_k.$$

Because the X_k are independent, zero-mean random variables, we know that if $n \neq k$, then $E(X_n X_k) = 0$. Thus, we see that the expected value of $|a_0|^2$ is

$$E(|a_0|^2) = NE(X_k^2). \tag{7.1}$$

We would like to examine the variance of $|a_0|^2$. First, consider $E(|a_0|^4)$. We find that

$$E(|a_0|^4) = NE(X_i^4) + 3N(N-1)E^2(X_i^2).$$

(See Exercise 5 for a proof of this result.) Thus, the variance of the measurement is

$$E(|a_0|^4) - E^2(|a_0|^2) = NE(X_i^4) + 2N^2 E^2(X_i^2) - 3NE^2(X_i^2)$$
$$= N\sigma_{X^2}^2 + 2(N^2 - N)E^2(X_i^2).$$

Clearly, the variance of $|a_0|^2$ is $O(N^2)$, and the standard deviation of $|a_0|^2$ is $O(N)$. That is, the standard deviation is of the same order as the measurement. This shows that taking larger values of N—taking more measurements—does not do much to reduce the uncertainty in our measurement of $|a_0|^2$. In fact, this problem exists for all the a_m, and it is also a problem when the measured values, X_k, are not IID random variables.

7.2 The Solution

We have seen that the standard deviation of our measurement is of the same order as the expected value of the measurement. Suppose that rather than taking one long measurement, one takes many smaller measurements. If the measurements are independent and one then averages the measurements, then the variance of the average will decrease with the number of measurements while the expected value will remain the same.

Given a sequence of samples of a random signal, $\{X_0, \ldots, X_{N-1}\}$, define the *periodograms*, P_m, associated with the sequence by

$$P_m \equiv \frac{1}{N} \left| \sum_{k=0}^{N-1} e^{-2\pi jkm/N} X_k \right|^2, \qquad m = 0, \dots, N-1.$$

The value of the periodogram is the square of the absolute value of the mth element of the DFT of the sequence divided by the number of elements in the sequence under consideration. The division by N removes the dependence that the size of the elements of the DFT would otherwise have on N—a dependence that is seen clearly in (7.1).

The solution to the problem of the non-decreasing variance of the estimates is to *average* many estimates of the same variable. In our case, it is convenient to average measurements of P_m, and this technique is known as *the method of averaged periodograms*.

Consider the MATLAB® program of Figure 7.1. In the program, MATLAB takes a set of 2^{12} uncorrelated random numbers that are uniformly distributed over $(-1/2, 1/2)$, and estimates the power spectral density of the "signal" by making use of the method of averaged periodograms. The output of the calculations is given in Figure 7.2. Note that the more sets the data were split into, the less "noisy" the spectrum looks. Note too that the number of elements in the spectrum decreases as we break up our data into smaller sets. This happens because the number of points in the DFT decreases as the number of points in the individual datasets decreases.

It is easy to see what value the measurements *ought* to be approaching. As the samples are uncorrelated, their spectrum ought to be uniform. From the fact that the MATLAB-generated measurements are uniformly distributed over $(-1/2, 1/2)$, it easy to see that

$$E(X_k^2) = \int_{-1/2}^{1/2} \alpha^2 \, d\alpha = \left. \frac{\alpha^3}{3} \right|_{-1/2}^{1/2} = \frac{1}{12} = 0.08\overline{3}.$$

Considering (7.1) and the definition of the periodogram, it is clear that the value of the averages of the 0th periodograms, P_0, ought to be tending to $1/12$. Considering Figure 7.2, we see that this is indeed what is happening—and the more sets the data are split into, the more clearly the value is visible. As the power should be uniformly distributed among the frequencies, *all* the averages should be tending to this value—and this too is seen in the figure.

7.3 Warm-up Experiment

MATLAB has a command that calculates the average of many measurements of the square of the coefficients of the DFT. The command is called psd (for power spectral density). (See [7] for more information about the power spectral density.) The format of the psd command is psd(X,NFFT,Fs,WINDOW) (but note that in MATLAB 7.4 this command is considered obsolete). Here, X is the data whose PSD one would like to find, NFFT is the number of points in each

```
% A simple program for examining the PSD of a set of
% uncorrelated numbers.
N = 2^12;
% The next command generates N samples of an uncorrelated random
% variable that is uniformly distributed on (0,1).
x = rand([1 N]);
% The next command makes the ''random variable'' zero-mean.
x = x - mean(x);

% The next commands estimate the PSD by simply using the FFT.
y0 = fft(x);
z0 = abs(y0).^2/N;

%The next commands break the data into two sets and averages the
%periodograms.
y11 = fft(x(1:N/2));
y12 = fft(x(N/2+1:N));
z1 = ((abs(y11).^2/(N/2)) + (abs(y12).^2/(N/2)))/2;

%The next commands break the data into four sets and averages the
%periodograms.
y21 = fft(x(1:N/4));
y22 = fft(x(N/4+1:N/2));
y23 = fft(x(N/2+1:3*N/4));
y24 = fft(x(3*N/4+1:N));
z2 = (abs(y21).^2/(N/4)) + (abs(y22).^2/(N/4));
z2 = z2 + (abs(y23).^2/(N/4)) + (abs(y24).^2/(N/4));
z2 = z2 / 4;

%The next commands break the data into eight sets and averages the
%periodograms.
y31 = fft(x(1:N/8));
y32 = fft(x(N/8+1:N/4));
y33 = fft(x(N/4+1:3*N/8));
y34 = fft(x(3*N/8+1:N/2));
y35 = fft(x(N/2+1:5*N/8));
y36 = fft(x(5*N/8+1:3*N/4));
y37 = fft(x(3*N/4+1:7*N/8));
y38 = fft(x(7*N/8+1:N));
z3 = (abs(y31).^2/(N/8)) + (abs(y32).^2/(N/8));
z3 = z3 + (abs(y33).^2/(N/8)) + (abs(y34).^2/(N/8));
z3 = z3 + (abs(y35).^2/(N/8)) + (abs(y36).^2/(N/8));
z3 = z3 + (abs(y37).^2/(N/8)) + (abs(y38).^2/(N/8));
z3 = z3 / 8;
```

Fig. 7.1. The MATLAB program

```
%The next commands generate the program's output.
subplot(4,1,1)
plot(z0)
title('One Set')
subplot(4,1,2)
plot(z1)
title('Two Sets')
subplot(4,1,3)
plot(z2)
title('Four Sets')
subplot(4,1,4)
plot(z3)
title('Eight Sets')
print -deps avg_per.eps
```

Fig. 7.1. The MATLAB program (continued)

FFT, Fs is the sampling frequency (and is used to normalize the frequency axis of the plot that is drawn), and WINDOW is the type of window to use. If WINDOW is a number, then a Hanning window of that length is used. Use the MATLAB help command for more details about the psd command.

Use the MATLAB rand command to generate 2^{16} random numbers. In order to remove the large DC component from the random numbers, subtract the average value of the numbers generated from each of the numbers generated. Calculate the PSD of the sequence using various values of NFFT. What differences do you notice? What similarities are there?

7.4 The Experiment

Note that as two ADuC841 boards are used in this experiment, *it may be necessary to work in larger groups than usual.*

Write a program to upload samples from the ADuC841 and calculate their PSD. You may make use of the MATLAB psd command and the program you wrote for the experiment in Chapter 4. This takes care of half of the system.

For the other half of the system, make use of the noise generator implemented in Chapter 6. This generator will be your source of random noise and is most of the second half of the system.

Connect the output of the signal generator to the input of the system that uploads values to MATLAB. Look at the PSD produced by MATLAB. Why does it have such a large DC component? Avoid the DC component by not plotting the first few frequencies of the PSD. Now what sort of graph do you get? Does this agree with what you expect to see from white noise?

Finally, connect a simple RC low-pass filter from the DAC of the signal generator to ground, and connect the filter's output to the A/D of the board

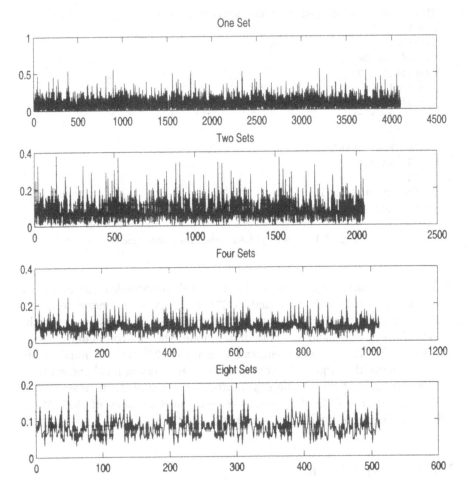

Fig. 7.2. The output of the MATLAB program when examining several different estimates of the spectrum

that uploads data to MATLAB. Observe the PSD of the output of the filter. Does it agree with what one expects? Please explain carefully.

Note that you may need to upload more than 512 samples to MATLAB so as to be able to average more measurements and have less variability in the measured PSD. Estimate the PSD using 32, 64, and 128 elements per window. (That is, change the NFFT parameter of the pdf command.) What effect do these changes have on the PSD's plot?

7.5 Exercises

1. What kind of noise does the MATLAB **rand** command produce? How might one go about producing true normally distributed noise?

2. (This problem reviews material related to the PSD.) Suppose that one passes white noise, $N(t)$, whose PSD is $S_{NN}(f) = \sigma_N^2$ through a filter whose transfer function is

$$H(f) = \frac{1}{2\pi j f \tau + 1}.$$

Let the output of the filter be denoted by $Y(t)$. What is the PSD of the output, $S_{YY}(f)$? What is the autocorrelation of the output, $R_{YY}(\tau)$?

3. (This problem reviews material related to the PSD.) Let $H(f)$ be the frequency response of a simple R-L filter in which the voltage input to the filter, $V_{in}(t) = N(t)$, enters the filter at one end of the resistor, the other end of the resistor is connected to an inductor, and the second side of the inductor is grounded. The output of the filter, $Y(t)$, is taken to be the voltage at the point at which the resistor and the inductor are joined. (See Figure 7.3.)

   a) What is the frequency response of the filter in terms of the resistor's resistance, R, and the inductor's inductance, L?

   b) What kind of filter is being implemented?

   c) What is the PSD of the output of the filter, $S_{YY}(f)$, as a function of the PSD of the input to the filter, $S_{NN}(f)$?

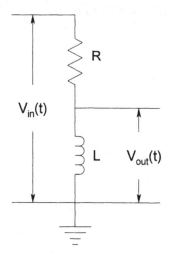

Fig. 7.3. A simple R-L filter

4. Using Simulink®, simulate a system whose transfer function is

$$H(s) = \frac{s}{s^2 + s + 10{,}000}.$$

Let the input to the system be band-limited white noise whose bandwidth is substantially larger than that of the filter. Use a "To Workspace" block to send the output of the filter to MATLAB. Use the PSD function to calculate the PSD of the output. Plot the PSD of the output against frequency. Show that the measured bandwidth of the output is in reasonable accord with what the theory predicts. (Remember that the PSD is proportional to the *power* at the given frequency, and *not* to the voltage.)

5. Let the random variables $X_0, \ldots, X_{N-1}$ be independent and zero-mean. Consider the product

$$(X_0 + \cdots + X_{N-1})(X_0 + \cdots + X_{N-1})(X_0 + \cdots + X_{N-1})(X_0 + \cdots + X_{N-1}).$$

a) Show that the only terms in this product that are not zero-mean are of the form X_k^4 or $X_k^2 X_n^2$, $n \neq k$.

b) Note that in expanding the product, each term of the form X_k^4 appears only once.

c) Using combinatorial arguments, show that each term of the form $X_k^2 X_n^2$ appears $\binom{4}{2}$ times.

d) Combine the above results to conclude that (as long as the samples are real)

$$E(|a_0|^4) = NE(X_k^4) + 6\frac{N(N-1)}{2}E^2(X_k^2).$$

Analog to Digital and Digital to Analog
Converters

8

The General Structure of Sampled-data Systems

Summary. In this chapter we consider the general structure of sampled-data systems. We will see that in sampled-data systems—whether they are used in the analysis of discrete-time signals as in Part I or to implement digital filters as in Part III—one must sample an analog signal, and often one must convert digitized samples of a signal back into an analog signal. Understanding how these conversions are made is the purpose of this part of the book—of Part II.

Keywords. sampled-data systems, analog to digital converters, digital to analog converters.

8.1 Systems for Spectral Analysis

When using digital techniques to analyze signals, one generally uses a system like that of Figure 8.1. The first element of the system takes the analog signal to be analyzed and converts it into a sequence of digital values; such elements are known as analog to digital converters. Over the next several chapters, we discuss several such systems.

Fig. 8.1. A typical system used in the spectral analysis of signals

8.2 Systems for Implementing Digital Filters

In Part III, we describe how one implements digital filters—filters that are implemented using a computer or microprocessor. Such systems generally have an analog input that must be converted into a stream of discretized values. In addition, the output of the digital system—the computer, microprocessor, or microcontroller—must be converted from a stream of discretized values into an analog signal. A sample system is shown in Figure 8.2. The element that converts a digital value into an analog signal—into a voltage—is called a digital to analog converter.

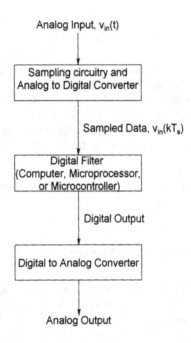

Fig. 8.2. A typical system used in the digital filtering of signals

9

The Operational Amplifier: An Overview

Summary. Before describing some of the converters that are available, we briefly consider the properties of the *operational amplifier*—the op-amp. The op-amp is a building block in many of the circuits described in the following chapters.

Keywords. operational amplifier, op-amp, buffer, inverting amplifier, non-inverting amplifier.

9.1 Introduction

The operational amplifier (or *op-amp*) is a circuit element that amplifies the difference of the voltages of its inputs. (For more information about op-amps and their history, the interested reader may wish to consult [12].) Let the two voltage inputs to the op-amp be denoted by V_+ and V_-. Then a first approximation to the output of the op-amp is $V_{\text{out}} \approx A(V_+ - V_-)$ where A, the amplification, is assumed to be a very large number. The standard symbol for an op-amp is given in Figure 9.1. An op-amp's input impedance is generally very large, and its output impedance is generally very small.

9.2 The Unity-gain Buffer

It is easy to make a unity-gain *buffer* using an op-amp. A unity-gain buffer is a circuit whose output voltage (ideally) equals its input voltage, whose input impedance is very large, and whose output impedance is low. A unity-gain buffer is often used to "transfer" a voltage from one subsystem to another without loading—without affecting—the subsystem from which the signal originates.

Consider the circuit of Figure 9.2. Making use of our first approximation to the op-amp, we find that

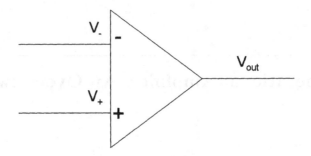

Fig. 9.1. The standard symbol for the operational amplifier

$$V_{out} = A(V_{in} - V_{out}) = AV_{in} - AV_{out}.$$

Solving for V_{out}, we find that

$$V_{out} = \frac{A}{A+1}V_{in}.$$

As long as $A \gg 1$, it is clear that $V_{out} \approx V_{in}$. Because of the op-amp's large input impedance, there is essentially no current flowing from the source of V_{in} into the op-amp. Because of the op-amp's low output impedance, the output of the op-amp can source a reasonable amount of current without the output voltage deviating from its anticipated value.

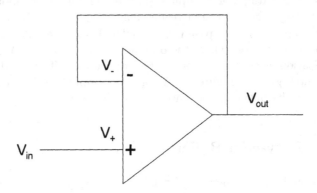

Fig. 9.2. A unity-gain buffer

9.3 Why the Signal is Fed Back to V_-

Let us consider a slightly more detailed model of the op-amp. Rather than assuming that the op-amp instantaneously multiplies the difference of its inputs by A, let us assume that (in the Laplace—the s—domain)

$$V_{\text{out}}(s) = \frac{A}{s\tau + 1}(V_+(s) - V_-(s)).$$

This associates a time constant, τ, with the op-amp and "means" that the op-amp does not react instantaneously to changes in its inputs. Using this model, we find that the relationship between the input and the output of the circuit of Figure 9.2 is

$$V_{\text{out}}(s) = \frac{A}{s\tau + 1}(V_{\text{in}}(s) - V_{\text{out}}(s)) \Rightarrow V_{\text{out}}(S) = \frac{A}{s\tau + 1 + A}V_{\text{in}}(s).$$

The transfer function of the unity-gain buffer is

$$T(s) = \frac{A}{s\tau + 1 + A}.$$

This transfer function has a single pole at $s = (-1 - A)/\tau$. As this pole is in the left half-plane, the buffer is stable [6].

Consider the system of Figure 9.3. We find that

$$V_{\text{out}}(s) = \frac{A}{s\tau + 1}(V_{\text{out}}(s) - V_{\text{in}}(s)) \Rightarrow V_{\text{out}}(S) = \frac{-A}{s\tau + 1 - A}V_{\text{in}}(s).$$

Here, the lone pole is located at $(-1+A)/\tau$. As long as $A > 1$, something that is true for all op-amps, this system is *unstable*. Generally speaking, in order for an op-amp-based system to be stable, the feedback *must* be connected to V_-—to the inverting input.

9.4 The "Golden Rules"

By making use of two simple "golden" rules (that are simplifications of how op-amps really behave), one can often "intuit" how an op-amp circuit will behave. The first rule is that when an op-amp circuit that employs feedback is operating correctly, the difference in voltage between the two input terminals of the op-amp, the difference between V_+ and V_-, is negligible. The second rule is that (to a good approximation) no current enters (or leaves) the two inputs of the op-amp.

Looking back at Figure 9.2, we see that making use of the first rule, $V_{\text{in}} = V_+ = V_- = V_{\text{out}}$. Additionally, from the second rule it is clear that the input to the buffer draws no current—and cannot load the circuit that provides the input, V_{in}.

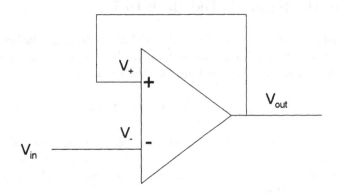

Fig. 9.3. An unstable unity-gain buffer

9.5 The Inverting Amplifier

Consider the circuit of Figure 9.4. Let us use the golden rules to examine what
the circuit does. First of all, because the inverting (V_-) and non-inverting (V_+)
inputs are assumed to be at the same voltage, the voltage at the non-inverting
input must be $0\,\mathrm{V}$. One says that there is a *virtual ground* at the inverting
input. This being the case, the current flowing in the resistor $R1$ must be
$V_{\mathrm{in}}/R1$ amperes. Because no current flows into an op-amp (according to the
second rule), the voltage drop across $R2$ must be $(V_{\mathrm{in}}/R1)R2$. As the drop
is from the voltage of the inverting input—which is approximately zero—the
output voltage must be

$$V_{\mathrm{out}} = -\frac{R2}{R1}V_{\mathrm{in}}.$$

This formula is valid for reasonably low-frequency signals. The circuit is called
an *inverting* amplifier because the sign of the output is always opposite to that
of the input.

9.6 Exercises

1. By using the golden rules, show that the circuit of Figure 9.3 "should be"
   (stability considerations aside) a unity-gain buffer.
2. Show that the circuit of Figure 9.5 is a non-inverting amplifier and that
   (at low frequencies)

$$V_{\mathrm{out}} = \left(1 + \frac{R2}{R1}\right)V_{\mathrm{in}}.$$

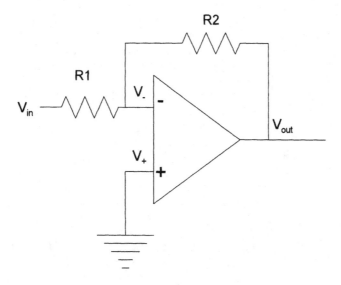

Fig. 9.4. An inverting amplifier

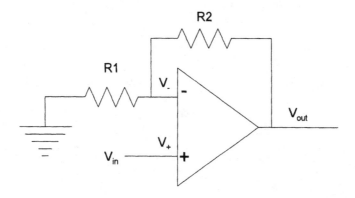

Fig. 9.5. A non-inverting amplifier

10

A Simple Digital to Analog Converter

Summary. There are many ways of implementing digital to analog converters (D/As or DACs)—circuits that take an input that consists of "ones" and "zeros" and convert the inputs into an analog output signal. (Often the "ones" are a reference voltage, V_{ref} and the "zeros" are $0\,V$.) In this chapter, we present one simple, albeit not very practical, DAC. This DAC can be used with many types of microprocessors and can be considered a purely passive DAC.

Keywords. digital to analog converter, DAC, passive DAC, buffer.

10.1 The Digital to Analog Converter

Consider the circuit of Figure 10.1. Let the inputs, b_0, b_1, and b_2 be either $0\,V$ or V_{ref}. Let us consider the output of the circuit when $b_2 = V_{ref}$, and let $b_1 = b_0 = 0\,V$. Then, we can redraw the circuit as shown in Figure 10.2.

Let us determine the output of the circuit in the second form. The resistance of the two resistors in parallel is

$$\text{Resistance} = \frac{1}{\frac{1}{2R} + \frac{1}{4R}} = \frac{4R}{3}.$$

As this resistor is in series with the R ohm resistor, the output of the circuit is

$$V_{out} = \frac{\frac{4R}{3}}{R + \frac{4R}{3}} V_{ref} = \frac{4}{7} V_{ref}.$$

Similar logic shows that if $b_0 = b_2 = 0, b_1 = V_{ref}$, then $V_{out} = (2/7)V_{ref}$, and if $b_1 = b_2 = 0, b_0 = V_{ref}$, then $V_{out} = (1/7)V_{ref}$.

The *principle of superposition* states that, given a linear circuit, if a set of inputs $\mathcal{I}_1$ leads to the output $\mathcal{O}_1$, and if a second set of inputs $\mathcal{I}_2$ leads to the output $\mathcal{O}_2$, then if one uses $a_1\mathcal{I}_1 + a_2\mathcal{I}_2$ as the input to the system, then the output of the system will be $a_1\mathcal{O}_1 + a_2\mathcal{O}_2$. The principle of superposition

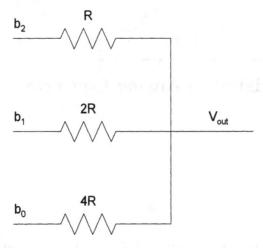

Fig. 10.1. A very simple DAC

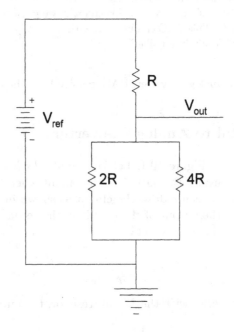

Fig. 10.2. The circuit redrawn

makes it relatively simple to determine the output of the system with any set of inputs. To determine the output for a generic input, one need only take the combination of the inputs whose sum gives the desired input. For example, to find the output when $b_0 = b_1 = b_2 = V_{\text{ref}}$, sum the outputs of all three of the previously calculated cases. This gives V_{ref}. Thus, when all three bits are "high," the output is V_{ref}.

It is easy to show that the output of the circuit is

$$V_{\text{out}} = \frac{b_0 + 2b_1 + 4b_2}{7}.$$

Thus, the output of the circuit is proportional to the binary value of the number $(b_2 b_1 b_0)_{\text{base 2}}$ (where the digit b_i is thought of as a one if $b_i = V_{\text{ref}}$ and is thought of as a zero if $b_i = 0\,\text{V}$). One can extend this circuit to an N-bit DAC by continuing the pattern of inputs and resistors. See Figure 10.3.

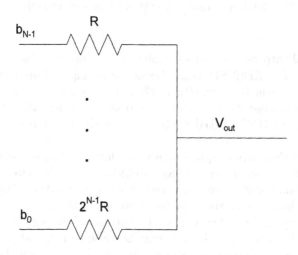

Fig. 10.3. A very simple N-bit DAC

10.2 Practical Difficulties

The first problem with this DAC is that if one loads its output—if one draws current from its output—one will "unbalance" the circuit. The loading causes the circuit's output voltage to change, and the circuit will no longer accurately convert our digital signal to a proportional analog signal. To prevent the next stage from loading the DAC, it is a good idea to add a unity-gain buffer to the output of the DAC. The improved circuit is shown in Figure 10.4.

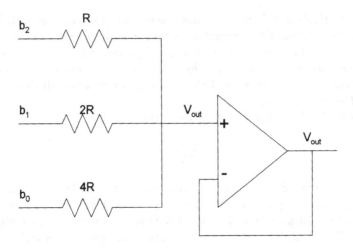

Fig. 10.4. The three-bit DAC with an output buffer

The digital outputs of a microcontroller are not meant to be used to source much current. The ADuC841's datasheets, for example, claim that the maximum output current is about $80\,\mu A$. That is a problem as that would make the maximum voltage drop across a $10\,K$ resistor no more than $800\,mV$. (In a single trial, the DAC worked fairly well. Clearly, $80\,\mu A$ is not always the actual limit.)

A possible solution to the problem is to use larger resistors—perhaps $100\,K$, $200\,K$, and $400\,K$ resistors. A practical problem with such a circuit is that as there is a certain capacitance associated with the pins of the I/O ports, large resistors may increase the rise time to an unacceptable level.

Another practical problem is that the outputs of the digital I/O pins are not generally $0\,V$ and V_{ref}. The zero may be close to $1\,V$, and the pin's high output will generally be similarly distant from its nominal value. DACs of this type will probably not be very accurate unless steps are taken to regulate the voltage of the digital inputs.

10.3 The Experiment

Write a very short program to make the output of the pins P3.0, P3.1, and P3.2 of the ADuC841 repeat the sequence 000, 001, 010, 011, 100, 101, 110, 111 indefinitely. Build the circuit of Figure 10.1, and let P3.0 be b_0, P3.1 be b_1, and P3.2 be b_2. Let $R = 5\,K$. Analyze the output of the circuit while running the program you wrote.

Next, implement the circuit of Figure 10.1 but with the $5\,K$ resistor replaced by a $100\,K$ resistor, the $10\,K$ resistor replaced by a $200\,K$ resistor and the $20\,K$ resistor replaced by a $400\,K$ resistor. Examine the output, describe how it differs from the output of the previous circuit, and explain *why* the differences you have observed exist. Please pay particular attention to the voltage levels and the apparent rise time of the output.

10.4 Exercises

1. What simple additions could one make to the circuit of Figure 10.1 to rid oneself of the problem of the circuit loading the ADuC841's digital outputs? (Please keep your answer very simple.)

11

The Binary Weighted DAC

Summary. In the previous chapter, we considered a simple, entirely passive DAC. In this chapter, we consider the binary weighted DAC. This DAC is also very simple but makes use of an op-amp in a way that is fundamental to the DAC's performance.

Keywords. binary weighted DAC, op-amp.

11.1 The General Theory

The circuit of Figure 11.1 shows an N-bit binary weighted DAC. The DAC's name comes from the way the resistors have been chosen; each resistor is twice as large as the preceding resistor. This is a very simple DAC that makes use of an op-amp in a way that is fundamental to the DAC's operation.

In Figure 11.1, the non-inverting input, V_+, is tied to ground. According to the golden rules, the voltage at the inverting input must be approximately $0\,\mathrm{V}$ as well. That is why there is a *dashed ground symbol* at the inverting input.

The current in the branch to which b_i is an input will be $b_i/(2^{(N-1)-i}R)$. Because (according to the golden rules) no current enters the op-amp, the total current passing through the R ohm resistor sitting "above" the op-amp is

$$\text{total current} = \sum_{i=0}^{N-1} \frac{b_i}{2^{(N-1)-i}R} = \frac{1}{2^{N-1}R} \sum_{i=0}^{N-1} 2^i b_i.$$

The voltage at the output is the voltage that drops across the R ohm resistor is positioned above the op-amp. As this voltage drops from zero—from the virtual ground—we find that the voltage at the output is

$$V_{\text{out}} = -\frac{1}{2^{N-1}} \sum_{i=0}^{N-1} 2^i b_i.$$

Clearly, this voltage is proportional to the binary value of the word $b_{N-1} \cdots b_0$.

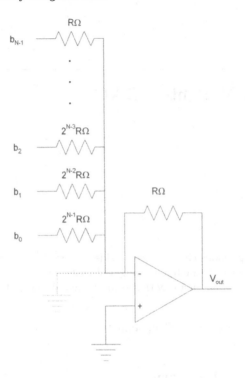

Fig. 11.1. An N-bit binary weighted DAC. The dashed portion is not physically present; the ground at the inverting input is a *virtual ground*.

This DAC is conceptually very simple, but it suffers from a serious short-coming. The resistors must have resistances of $2^k R$ ohm. If this proportionality is not maintained, then the output of the DAC will not be proportional to its input. Suppose, for example, that one has an eight-bit DAC all of whose resistors are in the exact proportions desired except that the value of the resistor connected to the input b_7 is 1% too small. When one translates the number 01111111 into a voltage, one finds that the associated output voltage is

$$V_{\text{out}} = -\frac{127}{128}V_{\text{ref}} = -0.9921875V_{\text{ref}}$$

where V_{ref} is the voltage input to b_i when it is associated with a logical one. When one translates the number 10000000 into a voltage, one finds that, because of the 1% error in the value of the resistor, the output voltage is

$$V_{\text{out}} = -\frac{128}{128}V_{\text{ref}}0.99 = -0.99V_{\text{ref}}.$$

We find that even though 10000000 *should* correspond to a larger voltage than the voltage which corresponds to 01111111, it does not. A 1% error in one of the resistor values led to our "losing" one bit of accuracy.

Because of the need for accuracy in the various resistor values called for by the binary weighted DAC, such DACs generally have eight or fewer bits of accuracy.

11.2 Exercises

1. What accuracy is required of the resistor connected to b_{15} in a 16-bit binary weighted DAC if the 16th bit, b_0, is to be meaningful?

12

The R-2R Ladder DAC

Summary. The R-2R ladder DAC is another passive DAC. The R-2R ladder DAC makes use of only two resistor values—but it requires $2N$ resistors to implement an N-bit converter.

Keywords. R-2R ladder, passive DAC, buffer, superposition.

12.1 Introduction

A three-bit example of an R-2R ladder DAC is given in Figure 12.1. As we will show, the output of the DAC is

$$V_{\text{out}} = \frac{4b_2 + 2b_1 + b_0}{8}. \tag{12.1}$$

One can extend the circuit to any number of bits by continuing the pattern of R and $2R$ ohm resistors.

12.2 The Derivation

We consider the effect of any one bit being high, or, in other words, of a given digital input being equal to V_{ref}, while the rest are low—are held at $0\,\text{V}$. Then, we make use of the principle of superposition to arrive at (12.1). First consider the output value for the input word 100_2. In terms of voltages, we have $b_0 = b_1 = 0$ and $b_2 = V_{\text{ref}}$. Our circuit is then equivalent to the circuit of Figure 12.2. It is clear that the leftmost two resistors are equivalent to one R ohm resistor. This is in series with the next R ohm resistor. Combined, the leftmost three resistors are a $2R$ ohm resistor in parallel with the next $2R$ ohm resistor. Thus, one finds that the leftmost four resistors are equivalent to a single R ohm resistor. Combining this with the next R ohm resistor, we find

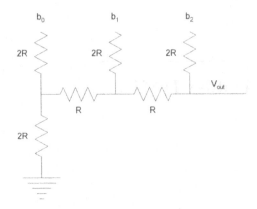

Fig. 12.1. A three-bit R-2R ladder DAC

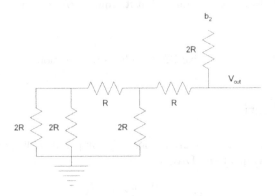

Fig. 12.2. An equivalent circuit when $b_0 = b_1 = 0$ and $b_2 = V_{\text{ref}}$

that to the rest of the circuit the leftmost five resistors act as a single $2R$ ohm resistor. One finds that the circuit's output, V_{out}, is $b_2/2$.

What happens when $b_1 = V_{\text{ref}}$ and $b_0 = b_2 = 0$? As we have seen, in this case the leftmost three resistors are equivalent to a single $2R$ ohm resistor. The equivalent circuit here is given by Figure 12.3. Considering the circuit in the dashed box as a two port and applying Thévenin's theorem [18], we find that the circuit can be replaced by a $b_1/2$ volt voltage source and an R ohm resistor in series; the equivalent circuit is given in Figure 12.4. It is now clear that

$$V_{\text{out}} = b_1/4.$$

It is left as an exercise (see Exercise 1) to show that, when $b_0 = V_{\text{ref}}$ and $b_1 = b_2 = 0$, $V_{\text{out}} = b_0/8$. The principle of superposition allows us to conclude that, in general,

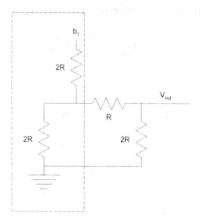

Fig. 12.3. An equivalent circuit when $b_0 = b_2 = 0$ and $b_1 = V_{\text{ref}}$

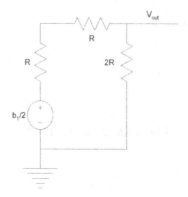

Fig. 12.4. The equivalent circuit after making use of Thévinin's theorem

$$V_{\text{out}} = \frac{4b_2 + 2b_1 + b_0}{8}.$$

The output of the DAC is proportional to the value of the binary number $b_2 b_1 b_0$.

12.3 Exercises

1. Show that when $b_0 = V_{\text{ref}}$ and $b_1 = b_2 = 0$, the output of the R-2R DAC is $b_0/8$.
2. Explain what should be added to the output of the circuit of Figure 12.1 in order to prevent the circuit's output from being loaded by the next stage.

3. Prove that the output of the circuit of Figure 12.5 is related to its inputs by the formula

$$V_{\text{out}} = \frac{2b_1 + b_0}{4}.$$

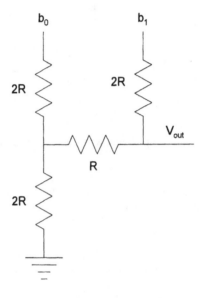

Fig. 12.5. A two-bit DAC

13

The Successive Approximation Analog to Digital Converter

Summary. Having seen several types of DACs, we move on to analog to digital converters. In this chapter, we describe and analyze an analog to digital converter that contains a DAC and control logic; we consider the successive approximation analog to digital converter.

Keywords. analog to digital converter, successive approximation ADC, sample-and-hold.

13.1 General Theory

There are many systems that convert analog input signals into digital output signals. These systems are known as analog to digital converters (A/Ds, ADCs, or, colloquially, "A to Ds"). The successive approximation A/D is built around a control unit, a DAC, and a comparator. See Figure 13.1.

The principles of operation of the successive approximation ADC are quite simple. The output of the DAC is constantly compared to the voltage input to the ADC. The control logic searches for the voltage that is nearest to the input voltage. This search can be carried out in many different ways. The simplest search method is to start with a digital input of zero and to increase the digital input by one unit until the comparator changes state. At that point, the controller knows that the output of the DAC has just passed the correct voltage. It stops adding one to the binary value of the input to the DAC and outputs the appropriate digital value—the value currently being input to the DAC.

This method of searching for the correct voltage is clearly inefficient. In the worst case scenario, it will take 2^N steps to reach the correct voltage. A better way to search is to use a binary search. The controller starts with the input to the DAC being $100 \cdots 0$—which corresponds to half the maximum output of the DAC. The controller examines the output of the comparator to see whether the output of the DAC is greater than, or less than, the input. If

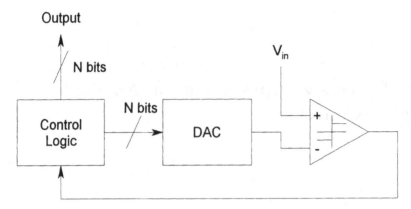

Fig. 13.1. A schematic representation of a successive approximation ADC

it is greater than the input, then the controller fixes the first bit of the output of the converter at zero and reduces the output of the DAC to one quarter its maximum output. If the output of the DAC is less than the input to the ADC, then the controller fixes the first bit of the output of the converter at one and increases the output of the DAC to three quarters of its maximum. In the next cycle, the controller works on determining the next lower bit of the output of the ADC. This process continues until the controller has determined the value of each of the bits of the output. In this way, one gains one bit of accuracy each cycle, and the search requires N cycles.

13.2 An Example

Consider a successive approximation ADC that has an internal three-bit binary weighted DAC with $V_{ref} = 2.5$ V. Let the DAC's output voltage increase in steps of $(2.5/8)$ V to a maximum of $(7/8)$ 2.5 V, and let the input to the ADC be 1.7 V. Initially, the control logic of the ADC outputs the binary word 100 to the DAC. The DAC outputs $(4/8)2.5 = 1.25$ V. As this voltage is lower than the voltage being input, the control logic fixes the first bit of the output at 1 and increases the voltage used in the comparison stage by changing the input to the DAC to 110. This causes the output of the DAC to change to $(6/8)2.5 = 1.875$ V. As this value is too large, the control logic fixes the second bit of the output at zero and halves the amount previously added to 1.25 V by changing the value output to the DAC to 101. This causes the voltage being compared to the input to change to $(5/8)2.5 = 1.5625$ V. As this value is less than the value of the input voltage, the control logic determines that the final output of the ADC is 101.

13.3 The Sample-and-hold Subsystem

As described above, a successive approximation A/D compares the input volt-age to a reference voltage over a relatively long period of time. It is generally important that the input to the A/D be forced to remain constant during the interval in which the comparisons are being carried out. A schematic diagram of a subsystem that samples and holds an input value is given in Figure 13.2. Such subsystems are know as *sample-and-hold* circuits. When switch S1 is open, the value on the capacitor should remain constant. When the switch is closed, the voltage on the capacitor becomes equal to the voltage currently at the input. By closing the switch briefly, one *samples* the input voltage, and by opening the switch one *holds* the previously sampled voltage. (This cir-cuit is a semi-practical implementation of the system whose properties were discussed in Section 1.1.) When used with an A/D, the A/D must be forced *not* to process the input until the output of the sample-and hold circuit has converged to the value of its input.

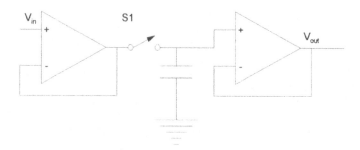

Fig. 13.2. A simple sample-and-hold circuit

14

The Single- and Dual-slope Analog to Digital Converters

Summary. We now consider the single-slope and the dual-slope ADCs. Both ADCs make use of simple op-amp circuits and control logic to do most of their work. We explain why the slightly more complicated dual-slope ADC is generally a better choice of ADC than the single-slope converter.

Keywords. single-slope converter, dual-slope converter, sensitivity to parameter values.

14.1 The Single-slope Converter

Consider the circuit of Figure 14.1. As the op-amp's non-inverting input, V_+, is tied to ground, by making use of the golden rules we find that there is a virtual ground at the inverting input. As long as the circuit's input is connected to V_{ref}, the current in the resistor is V_{ref}/R. This being the case the charge on the capacitor is increasing linearly. Assuming that the charge on the capacitor is initially zero, the voltage on the capacitor is

$$V_{cap} = -\frac{V_{ref}}{RC}t. \tag{14.1}$$

(The minus sign is here because the voltage *drops* across the capacitor, and the side of the capacitor through which the current "enters" is tied to the virtual ground.)

We find that the capacitor's voltage decreases linearly with time. To make this system the core of a converter, we need two more parts—a microprocessor and a comparator. The converter works as follows. At time zero, the microprocessor causes the voltage at the input to the op-amp circuit to go from 0 V to V_{ref}. The microprocessor keeps track of how much time has passed since the input went high. Every T_s seconds, the microprocessor checks the output of the comparator. When the comparator's output changes, the microprocessor

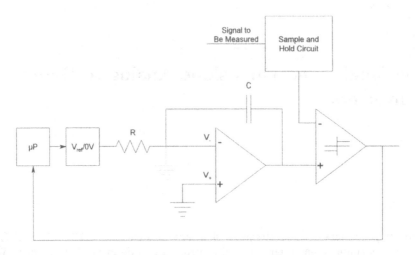

Fig. 14.1. The single-slope A/D. The block whose label is $V_{ref}/0\,V$ connects either V_{ref} or $0\,V$ to the resistor. The choice is controlled by the microprocessor.

"knows" that the output of the op-amp has exceeded the input—but barely. It then stores the number of sample periods, n, that it took the output of the op-amp circuit to exceed the value of the signal to be measured. The voltage at the output of the op-amp circuit—which approximates the voltage of the signal being measured—will be the voltage on the capacitor after n periods (of duration T_s) have gone by. From (14.1), it is clear that this voltage is

$$V_{\text{signal being measured}} \approx -\frac{V_{ref}T_s}{RC}n.$$

Thus, n is proportional to the voltage being measured, and n is the digital output of the A/D.

14.2 Problems with the Single-slope A/D

As presented, the single-slope A/D can only measure voltages whose sign differs from that of the reference voltage. This problem can be dealt with, if necessary.

A second problem has to do with the sensitivity of our estimate to changes in the values of R and C. A microprocessor's time-base is generally quite accurate, so T_s is generally known quite accurately. The values of resistors and capacitors are more problematic. It would be very nice to develop an A/D whose accuracy was not tied to the accuracy with which the values of R and C are known. The next type of A/D takes care of this problem.

14.3 The Dual-slope A/D

Consider the circuit of Figure 14.2. The circuit is very similar to the single-slope A/D. There are two major differences

- the input to the op-amp circuit can be either V_{ref} or the signal to be measured, and
- the comparator checks when the output of the op-amp circuit returns to zero.

In this circuit, the microprocessor initially sets the input to the op-amp circuit to the voltage to be measured for n clock cycles. It then switches the input to the op-amp circuit to V_{ref}—which must be opposite in polarity to the voltage to be measured. It maintains this voltage until the cycle in which the comparator determines that the sign of the output of the op-amp circuit has changed. Suppose that this take m cycles from the point at which the microprocessor switches the voltage source. Following the logic used in the case of the single-slope converter, we find that

$$\frac{V_{in}}{RC}T_s n + \frac{V_{ref}}{RC}T_s m \approx 0.$$

Solving for V_{in}, we find that

$$V_{in} \approx \frac{-V_{ref}}{n}m.$$

We find that V_{in} is proportional to m. Thus, m is the digital version of the voltage V_{in}. The beauty of this system is that the constant of proportionality is no longer a function of R or C. We can now use less expensive resistors and capacitors in our design, and we no longer need to worry (much) about the effects of aging or temperature on the values of R or C.

14.4 A Simple Example

Let $T_s = 100\,\mu s$, $n = 10$, and let $V_{ref} = -2.5\,V$. Suppose that $R = 100\,K$ and $C = 10\,nF$, and let $V_{in} = 1.1\,V$. Assuming that at time $t = 0$—when the measurement begins—the capacitor is fully discharged, during the first millisecond the voltage at the output of the op-amp will be

$$V_C = \frac{-V_{in}}{RC}t = -1,100t.$$

From this point on, the output will be

$$V_C = \frac{-V_{ref}}{RC}(t - 0.001) - 1.1 = 2,500(t - 0.001) - 1.1.$$

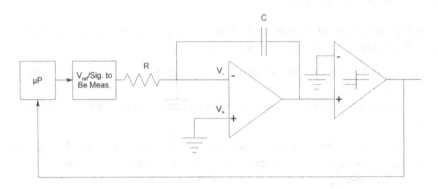

Fig. 14.2. The dual-slope A/D

(See Figure 14.3 to see the output of the "op-amp" of a Simulink model of this A/D.) It is easy to see that V_C goes from negative to positive when $t = 1.44$ ms. Thus, the final value of m will be 5. This corresponds to an estimate of

$$V_{in} \approx \frac{2.5}{10} 5 = 1.25 \, \text{V}.$$

14.5 Exercises

1. Repeat the calculations of Section 14.4, but let $n = 20$. What is the estimated value of V_{in} at the end of the conversion?

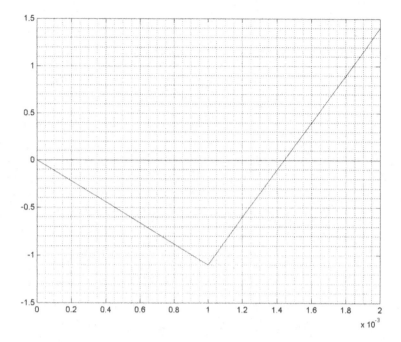

Fig. 14.3. The op-amp's output as a function of time

15

The Pipelined A/D

Summary. We now consider the pipelined A/D. This A/D is constructed by "stringing together" several identical subunits. The pipelined A/D generally has a high throughput—it processes a lot of analog data quickly. The pipelined A/D suffers from processing delay, however.

Keywords. pipelined A/D, latency, accuracy, processing delay.

15.1 Introduction

If one would like an A/D that is capable of reasonably high accuracy, is reasonably fast, and is not too expensive, a pipelined A/D may be a good choice. A pipelined A/D is composed of many identical units like the unit shown in Figure 15.1.

The first stage in the unit is a sample-and-hold circuit. The next stage is an analog to digital converter. The output bit(s) of this converter become some of the bits in the final output of the converter. Next, a D/A reconverts the digital word to an analog voltage. (The levels of the A/D and the D/A are designed to be identical.) The next stage is to calculate the difference of the output of the D/A and the input voltage. The difference of the voltages is the "unconverted" part of the input voltage. The last stage rescales the output of the differencer so that its range is, once again, the full range of the A/D. This voltage contains the "unconverted" part of the signal, and by processing the rescaled "remainder" it is possible to make the conversion more accurate.

The idea that underlies the pipelined converter is that by cascading several blocks like that of Figure 15.1, one can make an arbitrarily precise A/D. Each block takes its input, converts several bits, and outputs a voltage that is proportional to the portion of its input that the block did not convert. The next block "sees" the output of the previous block—which is proportional to the unconverted part of the input—and extracts a few more bits from it. This process continues down the pipeline.

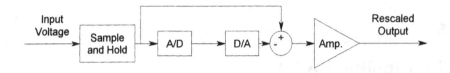

Fig. 15.1. A single unit of a pipelined A/D

15.2 The Fully Pipelined A/D

A fully pipelined A/D is a pipelined A/D in which each unit contributes a single bit to the output of the A/D. Consider the A/D of Figure 15.2. This figure shows a three-bit fully pipelined A/D. Let us consider a simple example of how the fully pipelined A/D operates.

Suppose that

- $V_{ref} = 2.5\,\mathrm{V}$,
- for voltages below 1.25 V, the A/D outputs a zero, and for voltages above this threshold, it outputs a one, and
- the DAC outputs 0 V when its input is a zero and 1.25 V when its input is a one.

Let the voltage input to the converter be 1.1 V, and assume that when we start, each of the sample-and-hold circuits holds 0 V. We find that at time zero all of the digital outputs are 0. After the first sample is taken, say at time 0^+, the first sample-and-hold holds 1.1 V and all of the bits remain 0. The output of the first amplifier is now $2 \cdot 1.1 = 2.2\,\mathrm{V}$. After the second clock cycle, the output of the second A/D is 1—as its input is greater than $V_{ref}/2$—and the output of the second amplifier is $2(2.2 - 1.25) = 1.9\,\mathrm{V}$. After the third clock cycle the output of the third A/D is 1 as well. The three bits given, 011, are the digital output of the converter. To convert from the digital output back to the (approximate) analog voltage at the input, one calculates

$$V_{\mathrm{in}} \approx \frac{4b_2 + 2b_1 + b_0}{8} V_{\mathrm{ref}}.$$

In making the conversion, the value of b_2 from the first cycle, the value of b_1 from the second cycle, and the value of b_0 from the third cycle are used. Until three cycles have passed, this circuit will not have completed a single conversion. Once three cycles have passed, however, this circuit outputs another conversion each cycle. Each finished conversion gives the digital value that corresponds to the input three cycles previously. If this delay—this *latency*—is acceptable, then the pipelined A/D is often a reasonable A/D to choose.

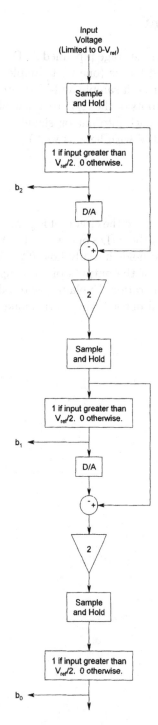

Fig. 15.2. A fully pipelined A/D

15.3 The Experiment

Using Simulink®, build a three-stage pipelined A/D converter. You may need to use two sample-and-hold blocks (one that samples on the rising edge and one on the falling edge) for each sample-and-hold circuit.

Connect the digital outputs of the A/D to a new block—a D/A—that will convert the digital signals back into analog signals. (Note that there will be synchronization issues to deal with here as well.)

15.4 Exercises

1. Let the values of the input to the circuit of Figure 15.2 be $\{1.0, 1.5, 2.3, 2.4\}$ at the first four cycles of the A/D's operation. Let $V_{ref} = 2.5\,\text{V}$, and assume that all the sample-and-holds initially hold $0\,\text{V}$. Find the values of b_2, b_1, and b_0 and the values of the outputs of the amplifiers for the first four periods of the A/D's operation. Explain how to take the calculated values and produce the digital outputs that correspond to the first and second analog inputs.

16
Resistor-chain Converters

Summary. Given a chain of identical resistors, it is conceptually simple to build either an A/D or a D/A. In this chapter, we describe the principles of operation of the resistor-chain DAC and the flash A/D.

Keywords. resistor-chain DAC, flash A/D, Karnaugh map, thermometer code.

16.1 Properties of the Resistor Chain

One way of deriving a set of voltages that are multiples of one another is to take a "chain" of identical resistors, to connect one side of the chain to a reference voltage, V_{ref}, and to ground the other side of the chain. See Figure 16.1. It is clear that the voltage between resistor i and $i+1$, V_i, is

$$V_i = \frac{i}{N} V_{\text{ref}}, \qquad i = 0, \ldots, N-1$$

where V_0 is the voltage between R_1 and ground.

16.2 The Resistor-chain DAC

In principle, it is simple to take a resistor chain and convert it into either a D/A or an A/D. Let us consider the D/A first. Suppose that the digital input to the converter is the n-bit binary word $b_{n-1} \cdots b_0$. If one takes the resistor chain and uses an analog multiplexer to connect $V_{2^{n-1}b_{n-1}+\cdots+2b_1+b_0}$ to the output, one has a conceptually simple D/A. In order to stop the load of the D/A from drawing current from the resistors and destroying the "balance" that creates the voltage divider, the output of the multiplexer should be fed into a buffer. A schematic drawing of such a resistor-chain D/A is given in Figure 16.2.

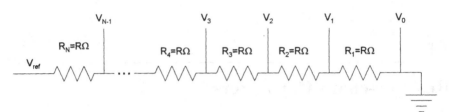

Fig. 16.1. A resistor chain. One end of the chain is connected to the reference voltage; the other end is grounded.

The resistor-chain DAC has one clear advantage: as the digital input increases, so does the analog output. No matter how poorly matched the supposedly identical resistors in the resistor chain are, as the digital input increases and the pointer—the analog multiplexer—moves to a resistor higher up in the resistor chain, the output voltage increases.

Building a 12-bit resistor-chain DAC is not something one would want to do by hand. It is not too hard to do when using standard chip manufacturing processes; DAC0 and DAC1 on the ADuC841 are such resistor-chain DACs.

16.3 The Flash A/D

Building an A/D using a resistor chain is also relatively straightforward. For an N-bit converter, one takes a resistor chain with 2^N resistors, one compares the voltage between each pair of resistors and the input voltage using a comparator, and one then feeds all $2^N - 1$ outputs to a decoder circuit. See Figure 16.3. Such an A/D is called a flash A/D. Because of the relative simplicity of this design, flash converters are generally very fast—as their name implies. Because of the way that the complexity (and hence price) of the circuit grows as the number of bits to be converted grows, flash A/Ds generally are *not used* when one needs more than eight bits in the final digital output.

For any given voltage, starting with the comparator nearest ground we find that the output of each comparator is one until we reach the first comparator for which the voltage on the resistor chain is higher than the input voltage. From that point onward, the output of the comparators is always zero. (This sort of code, where the data is contained in the point at which an output switches from one value to another, is called a *thermometer code*.) When designing the decoding circuitry, it is clear that there will be many "don't cares" in the truth table. This allows us to design relatively simple decoders. (See Exercise 2.)

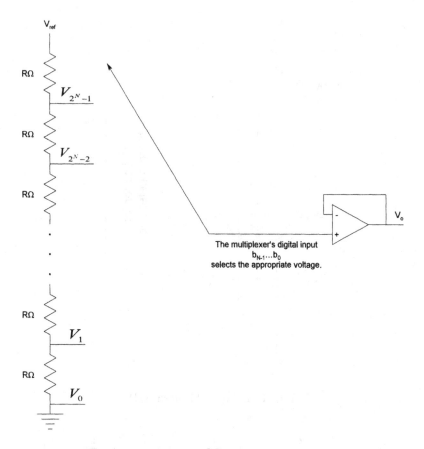

V_{ref}

$R\Omega$

V_{2^N-1}

$R\Omega$

V_{2^N-2}

$R\Omega$

The multiplexer's digital input
$b_{N-1}...b_0$
selects the appropriate voltage.

V_o

$R\Omega$

V_1

$R\Omega$

V_0

Fig. 16.2. A resistor-chain D/A. The unity-gain buffer at the output stops the voltage divider from being loaded by the DAC's load.

16.4 Exercises

1. Suppose that the resistors in the resistor chain of Figure 16.1 are not all precisely equal but rather that $R_i = R + \Delta R_i$ where the $\Delta R_i, i = 1, \ldots, 2^N$ are zero-mean random variables for which $\sigma^2_{\Delta R_i} = \epsilon^2 R^2$ (and where we let $\Delta R_0 \equiv 0$).

   a) Show that

   $$V_i/V_{\mathrm{ref}} = \frac{i/2^N + \frac{1}{R2^N}\sum_{k=0}^{i} \Delta R_k}{1 + \frac{1}{R2^N}\sum_{k=0}^{2^N} \Delta R_k}.$$

   b) Assuming that $\frac{1}{R2^N}\sum_{k=0}^{2^N} \Delta R_k \ll 1$, show that

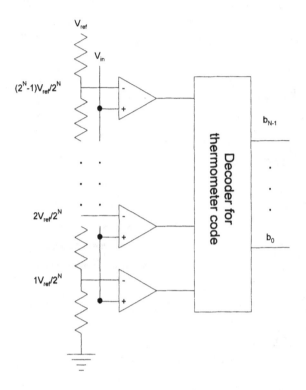

Fig. 16.3. An N-bit flash A/D

$$V_i/V_{\text{ref}} \approx \left(i/2^N + \frac{1}{R2^N} \sum_{k=0}^{i} \Delta R_k \right) \left(1 - \frac{1}{R2^N} \sum_{k=0}^{2^N} \Delta R_k \right)$$

$$\approx \frac{i}{2^N} + \frac{1}{R2^N} \sum_{k=0}^{i} \left(1 - \frac{i}{2^N} \right) \Delta R_k - \frac{i}{R2^{2N}} \sum_{k=i+1}^{2^N} \Delta R_k.$$

c) Now, calculate the root mean square (RMS) value of

$$\frac{1}{R2^N} \sum_{k=0}^{i} \left(1 - \frac{i}{2^N} \right) \Delta R_k - \frac{i}{R2^{2N}} \sum_{k=i+1}^{2^N} \Delta R_k.$$

You may assume that the random variables are independent.

2. A schematic diagram of a two-bit flash A/D is given in Figure 16.4. The truth table for b_0 and b_1 is given in Table 16.1. Please make use of Karnaugh maps to find the minimal realization of the two functions given in the table.

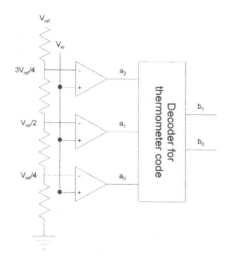

Fig. 16.4. A two-bit flash A/D

Table 16.1. The truth table for b_1 and b_0 as functions of a_0, a_1, and a_2. The letter X denotes a "don't care."

a_2	a_1	a_0	b_1	b_0
0	0	0	0	0
0	0	1	0	1
0	1	0	X	X
0	1	1	1	0
1	0	0	X	X
1	0	1	X	X
1	1	0	X	X
1	1	1	1	1

17

Sigma–Delta Converters

Summary. By making use of feedback in an interesting way, it is possible to design very accurate ADCs and DACs. In this chapter, we consider the sigma–delta A/D and the sigma–delta D/A. We find that though they can be very accurate, when they are most accurate they are also very slow.

Keywords. sigma–delta DAC, sigma–delta A/D, feedback, noise shaping, oversampling.

17.1 Introduction

We now consider sigma–delta (Σ–Δ) converters. Such converters make use of sums (from whence the Σ), differences (from whence the Δ), feedback, and low-pass filters in order to do their jobs. A generic sigma–delta converter is shown in Figure 17.1. This system can be considered a simplified model of an A/D *or* a D/A. Upon considering $V_{in}(s)$, $N(s)$, and $Y(s)$, we find that

$$Y(s) = N(s) + \frac{K}{s}(V_{in}(s) - Y(s)).$$

Solving for the output of the feedback system, $Y(s)$, we find that

$$Y(s) = \frac{s}{K+s}N(s) + \frac{K}{K+s}V_{in}(s). \tag{17.1}$$

What is *critical* here is that for small values of $s = j\omega$—for low frequencies—the noise term, $N(s)$, is attenuated greatly while the input is passed along almost without change. (Sigma–delta modulators are said to "shape" the noise spectrum and to have a *noise-shaping* property.) The low-pass filter through which $Y(s)$ passes should remove the noise component—which is largely located at high frequencies—without affecting the input, which is *assumed* to have a low-pass character.

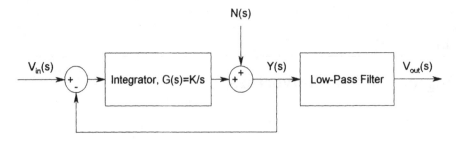

Fig. 17.1. A generic sigma–delta converter. $N(s)$ is a noise term serving to model the effect of the one-bit quantization used in sigma–delta converters.

17.2 The Sigma–Delta A/D

Consider the sigma–delta A/D of Figure 17.2. If we sample sufficiently quickly, then the sum that appears in Figure 17.2 satisfies

$$\sum_{k=0}^{N-1} e(k) = \frac{1}{T_s} \sum_{k=0}^{N-1} e(k)T_s \approx \frac{1}{T_s} \int_0^{NT_s} e(t)\, dt.$$

That is, the summer is (approximately) an integrator combined with a gain of $1/T_s$. Additionally, the comparator that follows the summer can be thought of as taking the output of the summer and "adding" quantization noise to it. Because of the very high sampling rate, the unit delay, a delay of one sampling period, can be considered an infinitesimal delay, and its effect on the system can be ignored. Thus our system looks like the generic system of Figure 17.1 where $K = 1/T_s$, and $N(s)$ is the quantization noise added by the comparator.

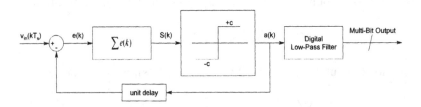

Fig. 17.2. A sigma–delta A/D. We assume that the input is already sampled, and that the system is clocked at the sample period, T_s. Note that the signals $e(k)$, $S(k)$, and $a(k)$ are continuous-time waveforms whose values change at sampling instants—at multiples of T_s. The summer samples its input once per sampling period.

Let us analyze the effect that the noise, $N(s)$, has on $Y(s)$ and $V_{out}(s)$. In order to proceed, we make two assumptions [1]:

1. The quantization noise is uniformly distributed between $-c$ and $+c$ (where $-c$ and $+c$ are output levels of the comparator of Figure 17.2).
2. The power spectral density of the quantization noise is constant from minus the Nyquist frequency, $-1/(2T_s)$, up to the Nyquist frequency, $1/(2T_s)$. It is zero outside of this interval.

The first assumption allows us to conclude that the mean (the expected value) of the quantization noise is zero, and its variance is

$$
\begin{aligned}
\sigma^2_{noise} &= E((n(t) - E(n(t))^2) \\
&= \int_{-\infty}^{\infty} (\alpha - 0)^2 f_{n(t)}(\alpha)\, d\alpha \\
&= \int_{-c}^{c} (\alpha - 0)^2 \frac{1}{2c}\, d\alpha \\
&= \frac{1}{c} \int_{0}^{c} \alpha^2\, d\alpha \\
&= c^2/3.
\end{aligned}
$$

As the variance of a random variable is equal to the integral of the random variable's power spectral density [7], we find that

$$
\int_{-1/(2T_s)}^{1/(2T_s)} S_{NN}(f)\, df = c^2/3.
$$

As $S_{NN}(f)$ is constant in the region over which we are integrating, we find that:

$$
S_{NN}(f) = \begin{cases} T_s c^2/3, & |f| < 1/(2T_s) \\ 0, & \text{elsewhere} \end{cases}.
$$

Consider the noise at the output of the comparator. Making use of (17.1), the linearity of the system, and the general theory of random signals and noise (see [7]), we find that the power spectral density of the noise at the output of the comparator is

$$
\text{PSD}(f) = \frac{(2\pi f)^2}{1/T_s^2 + (2\pi f)^2} T_s c^2/3
$$

for $|f| < 1/(2T_s)$, and it is zero elsewhere.

Now, consider the power spectral density at the output of the low-pass filter. Assume that the filter is an ideal unity-gain filter that passes frequencies up to F and removes all higher frequencies. We find that the power spectral density of the noise at the output, $S_{NN}(f)$, is

$$
S_{NN}(f) = \begin{cases} \frac{(2\pi f)^2}{1/T_s^2 + (2\pi f)^2} T_s c^2/3, & |f| < F \\ 0, & \text{otherwise} \end{cases}.
$$

Assuming that T_s and F are relatively small, the above function can be approximated by

$$S_{NN}(f) \approx \begin{cases} (2\pi f)^2 T_s^3 c^2/3, & |f| < F \\ 0, & \text{otherwise} \end{cases}.$$

To find the total noise power in the output, we need only integrate this function over the entire real axis. We find that the total noise power is

$$\text{total noise power} \approx 2(2\pi)^2(F^3/3)T_s^3 c^2/3 = (8\pi^2 c^2/9)(F/F_s)^3$$

where F_s is the sampling frequency (the reciprocal of the sampling period, T_s). As the noise power is the expected value of the square of the noise, the RMS value of the noise is

$$v_{\text{RMS}} = \sqrt{\text{total noise power}} \approx \sqrt{8/9}\pi c(F/F_s)^{3/2}.$$

How many bits of useful data will we have after we convert the output bitstream, $a(k)$, to a lower-rate stream of words? The maximum value at the output of the comparator is c. In order to obtain N bits of useful information when the input to the system is at its maximum, we must know that the noise value is less than $c/2^N$. Otherwise, the Nth bit will (essentially) correspond to the noise present—and not the signal. We find that

$$v_{\text{RMS}} \approx \sqrt{8/9}\pi c(F/F_s)^{3/2} \approx c/2^N$$

in order to develop an N-bit A/D. Taking logarithms, we find that the number of bits of accuracy we can expect from a sigma–delta A/D is

$$N \approx -(3/2)\log_2(F/F_s) + \text{constant}.$$

That is, *by doubling the sampling speed or halving the bandwidth of the converter, one gains one and a half bits of accuracy.*

17.3 Sigma–Delta A/Ds, Oversampling, and the Nyquist Criterion

In order for a sigma–delta converter to work properly, it is necessary that it sample very quickly—much faster than the Nyquist frequency for the signal at its input. The sigma–delta converter is said to *oversample* its input. Inputing a signal whose highest frequency is less than half the *actual* sampling rate of the converter will not lead to aliasing. Of course, the low-pass filter at the output of the converter will remove all signals whose frequency is greater than F, so the portions of the signal that are located at frequencies greater than F will be removed. They will not, however, lead to aliasing.

When using a sigma–delta A/D, it is often possible to employ a very simple anti-aliasing filter. The filter must pass frequencies up to F and must remove all frequencies about $F_s/2$. As $F_s/2 \gg F$, such a filter is generally easy to implement.

17.4 Sigma–Delta DACs

The principles of operation of the sigma–delta DAC are similar to those of the sigma–delta A/D. Figure 17.3 shows a schematic diagram of a simple sigma–delta DAC. Note how the analog elements of the sigma–delta A/D are digital in the sigma–delta DAC. The interested reader is referred to [8] for a detailed treatment of a simple sigma–delta DAC.

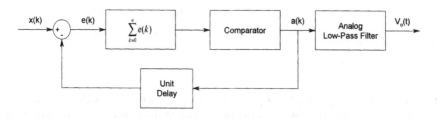

Fig. 17.3. A sigma–delta DAC

17.5 The Experiment

(This experiment requires some knowledge of discrete-time systems and may most profitably be performed after the material in Chapter 18 has been studied.)

Please use Simulink® to implement a sigma–delta ADC. Note that the transfer function of a summer is

$$T(z) = \frac{z}{z - 1},$$

and the transfer function of a unit delay is $1/z$.

Please design the model to calculate the mean square error of the converter. You may refer to the model of Figure 17.4 while building your system.

- After building the system, input signals whose frequencies are between 0 and F Hz. How well does the system track its inputs?
- What is the measured mean square error? Is the measurement in reasonable agreement with the theory developed in this chapter?
- Now, input signals whose frequencies are between F and $F_s/2$ Hz. What output do you see?
- Finally, input a signal whose frequency is slightly above $F_s/2$. What does the output look like now? Why is this output reasonable?

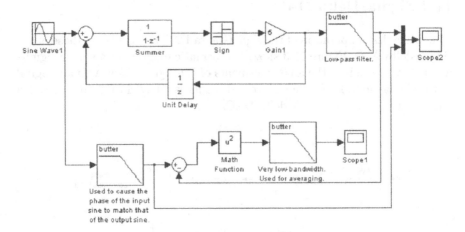

Fig. 17.4. A Simulink model of a sigma–delta A/D. The lower half of the system shown is used to calculate the noise at the output of the upper low-pass filter. Note that an analog low-pass filter is used in the converter. In principle, the filter should be a digital filter with digital output, but the analog filter makes "seeing" what is happening easier. The steady-state value seen on Scope1 is approximately the mean square noise voltage.

Part III

Digital Filters

18

Discrete-time Systems and the Z-transform

Summary. In engineering, most subject areas make use of specific mathematical tools. In the design and analysis of discrete-time systems, the most important of the mathematical tools is the Z-transform. In this chapter, we develop the Z-transform and its properties, and we show how to make use of the Z-transform to analyze discrete-time linear time-invariant systems. In the following chapters, we treat the problem of digital filter design.

Keywords. Z-transform, LTI system, region of convergence, sinusoidal steady state, stability.

18.1 The Definition of the Z-transform

Given a doubly infinite sequence of values $\{\ldots, a_{-1}, a_0, a_1, \ldots\}$, we define the two-sided (or *bilateral*) Z-transform of the sequence, $A(z)$, by the equation

$$A(z) \equiv \sum_{k=-\infty}^{\infty} a_k z^{-k}. \tag{18.1}$$

We generally use lowercase letters to represent sequences and uppercase letters to represent Z-transforms. As we see shortly, the Z-transform has many nice properties.

18.2 Properties of the Z-transform

18.2.1 The Region of Convergence (ROC)

The Z-transform of the sequence $\{a_k\}$ can be written

$$A(z) = \sum_{k=-\infty}^{\infty} a_k z^{-k} = \sum_{k=-\infty}^{-1} a_k z^{-k} + \sum_{k=0}^{\infty} a_k z^{-k}.$$

We consider the two halves of the sum that define $A(z)$ separately. Considering the sum

$$\sum_{k=0}^{\infty} a_k z^{-k},$$

we see that this sum *should* be well defined for z large enough. (When z is large, z^{-k} is small.) Suppose that the sum converges for some value $z = b$. Then we know that there exists a $C > 0$ for which $|a_k b^{-k}| < C$ for all k.

Making use of this result and the triangle inequality, we find that for any c for which $|c| > |b|$,

$$\left| \sum_{k=0}^{\infty} a_k c^{-k} \right| = \left| \sum_{k=0}^{\infty} a_k b^{-k} (c/b)^{-k} \right|$$

$$\leq \sum_{k=0}^{\infty} |a_k b^{-k}| |c/b|^{-k}$$

$$\leq C \sum_{k=0}^{\infty} |b/c|^k$$

$$\stackrel{\text{geometric series}}{=} C \frac{1}{1 - |b/c|}.$$

That is, if the sum converges at a point b, it converges for all c for which $|c| > |b|$.

Let us consider the other half of the sum

$$\sum_{k=-\infty}^{-1} a_k z^{-k}.$$

Supposing that the sum converges for some value b, it is easy to show that the sum must converge for all d that satisfy $|d| < |b|$. (See Exercise 2.)

Taking these two results, we find that the Z-transform converges in a region of the form

$$c_1 < |z| < c_2;$$

it does not converge for z such that $|z| < c_1$ or for which $|z| > c_2$. The Z-transform's behavior when $|z| = c_1$ or $|z| = c_2$ can be quite interesting. (See Exercise 7 for a nice example of such interesting behavior.) The region in which the Z-transform converges is known as the transform's *region of convergence* (or ROC).

Note that if $a_k = 0$ for all $k < 0$, then the ROC of the Z-transform contains a region of the form $|z| > c_1$. Also, from what we have already seen, it is clear that *inside* its ROC a Z-transform converges absolutely and uniformly. That is sufficient to allow us to interchange summation and integration and to interchange summation and differentiation in the *interior* of the ROC. (See [3] for more information about power series, their regions of convergence, and their properties.)

18.2.2 Linearity

The Z-transform is linear. That is, if one is given two sequences,

$$\{\ldots, a_{-1}, a_0, a_1, \ldots\}$$

and

$$\{\ldots, b_{-1}, b_0, b_1, \ldots\},$$

whose Z-transforms are $A(z)$ and $B(z)$, respectively, then the Z-transform of any linear combination of the sequences

$$\alpha\{\ldots, a_{-1}, a_0, a_1, \ldots\} + \beta\{\ldots, b_{-1}, b_0, b_1, \ldots\} \equiv$$
$$\{\ldots, \alpha a_{-1} + \beta b_{-1}, \alpha a_0 + \beta b_0, \alpha a_1 + \beta b_1, \ldots\}$$

is

$$\alpha A(z) + \beta B(z).$$

The region of convergence of the transform is (at least) the intersection of the ROCs of $A(z)$ and $B(z)$. For the proof, see Exercise 5.

18.2.3 Shifts

Suppose that one is given the sequence $\{\ldots, a_{-1}, a_0, a_1, \ldots\}$ whose Z-transform is $A(z)$, and one would like to calculate the Z-transform of the shifted sequence $\{\ldots, b_{-1}, b_0, b_1, \ldots\}$ where $b_k = a_{k-n}, n \in \mathcal{Z}$. One finds that

$$B(z) = \sum_{k=-\infty}^{\infty} b_k z^{-k}$$

$$= \sum_{k=-\infty}^{\infty} a_{k-n} z^{-k}$$

$$= z^{-n} \sum_{k=-\infty}^{\infty} a_{k-n} z^{-(k-n)}$$

$$= z^{-n} A(z).$$

The importance of this result cannot be overemphasized. The ROC associated with $B(z)$ will be the same as that associated with $A(z)$ (except, perhaps, for the points $z = 0$ and $z = \infty$).

18.2.4 Multiplication by k

Suppose that one is given the sequence $\{\ldots, a_{-1}, a_0, a_1, \ldots\}$ whose Z-transform is $A(z)$, and one would like to calculate the Z-transform associated with the sequence $\{\ldots, -1 \cdot a_{-1}, 0 \cdot a_0, 1 \cdot a_1, \ldots k a_k, \ldots\}$. Noting that

$$-z\frac{\mathrm{d}}{\mathrm{d}z}A(z) = -z\frac{\mathrm{d}}{\mathrm{d}z}\sum_{k=-\infty}^{\infty}a_k z^{-k}$$

$$= -z\sum_{k=-\infty}^{\infty}a_k(-k)z^{-k-1}$$

$$= \sum_{k=-\infty}^{\infty}a_k k z^{-k},$$

we find that the Z-transform of the new sequence is simply $-zA'(z)$. The ROC associated with this transform is the same as that associated with $A(z)$ (with the possible exception of the points on the boundary of the ROC).

18.3 Sample Transforms

18.3.1 The Transform of the Discrete-time Unit Step Function

Let us consider several simple examples. First, consider the discrete-time unit step function:

$$u_k \equiv \begin{cases} 0 & k < 0 \\ 1 & k \geq 0 \end{cases}.$$

The Z-transform of this sequence is

$$U(z) = \sum_{k=-\infty}^{\infty}u_k z^{-k} = \sum_{k=0}^{\infty}z^{-k}.$$

This sum is a geometric series whose ratio is z^{-1}. Thus, the sum is

$$A(z) = \frac{1}{1-z^{-1}} = \frac{z}{z-1}, \qquad |z| > 1.$$

We find that for this sequence, the ROC of the Z-transform is $|z| > 1$.

18.3.2 A Very Similar Transform

Next, consider the sequence

$$a_k = \begin{cases} -1 & k < 0 \\ 0 & k \geq 0 \end{cases}.$$

The Z-transform is

$$A(z) = \sum_{k=-\infty}^{\infty}a_k z^{-k} = -\sum_{k=-\infty}^{-1}z^{-k} = -\sum_{k=1}^{\infty}z^{k} = -z\sum_{k=0}^{\infty}z^{k}.$$

The last sum is a geometric series whose ratio is z. Thus, the sum is

$$A(z) = -z\frac{1}{1-z} = \frac{z}{z-1}, \qquad |z| < 1.$$

We find that for this sequence, the ROC is $|z| < 1$. *The only difference between the preceding two Z-transforms is the regions in which they converge.*

18.3.3 The Z-transforms of Two Important Sequences

Let us consider two important one-sided sequences. First, we consider the sequence

$$a_k = a^k u_k = \begin{cases} 0 & k < 0 \\ a^k & k \geq 0 \end{cases}.$$

The Z-transform of this sequence is

$$A(z) = \sum_{k=0}^{\infty} \alpha^k z^{-k}$$

$$= \sum_{k=0}^{\infty} (\alpha/z)^k$$

$$= \frac{1}{1 - \alpha/z}$$

$$= \frac{z}{z - \alpha}, \qquad |z| > |\alpha|.$$

Next, consider the sequence defined by

$$a_k = \sin(\omega k) u_k = \begin{cases} 0 & k < 0 \\ \sin(\omega k) & k \geq 0 \end{cases}.$$

Making use of the linearity of the Z-transform and the fact that

$$\sin(\omega k) = \frac{e^{j\omega k} - e^{-j\omega k}}{2j},$$

we find that the Z-transform of the sequence is

$$A(z) = \frac{1}{2j} \left(\frac{z}{z - e^{j\omega}} - \frac{z}{z - e^{-j\omega}} \right)$$

$$= \frac{1}{2j} \frac{z(e^{j\omega} - e^{-j\omega})}{z^2 - z(e^{j\omega} + e^{-j\omega}) + 1}$$

$$= \frac{z\sin(\omega)}{z^2 - 2z\cos(\omega) + 1}, \qquad |z| > 1.$$

18.3.4 A Two-sided Sequence

Consider the truly two-sided sequence

$$a_k = \alpha^{|k|}, \qquad |\alpha| < 1.$$

We find that

$$
\begin{aligned}
A(z) &= \sum_{k=-\infty}^{\infty} a_k z^{-k} \\
&= \sum_{k=-\infty}^{\infty} \alpha^{|k|} z^{-k} \\
&= \sum_{k=1}^{\infty} \alpha^k z^k + \sum_{k=0}^{\infty} \alpha^k z^{-k} \\
&= \alpha z \frac{1}{1-\alpha z} + \frac{1}{1-\alpha z^{-1}} \\
&= \frac{\alpha z}{1-\alpha z} + \frac{z}{z-\alpha} \\
&= \frac{z(1-\alpha^2)}{(1-\alpha z)(z-\alpha)}.
\end{aligned}
$$

The ROC is easily seen to be $|\alpha| < |z| < 1/|\alpha|$.

18.4 Linear Time-invariant Systems

Linear systems are those systems that obey the *principle of superposition*. Let $\{y_k\}$ be the output of a system when its input is some sequence $\{x_k\}$, and let $\{\tilde{y}_k\}$ be the system's output when its input is $\{\tilde{x}_k\}$. As we have seen (on p. 75), a system is said to satisfy the principle of superposition if for all such sequences the output of the sequence $\{ax_k + b\tilde{x}_k\}$ is $\{ay_k + b\tilde{y}_k\}$.

A system is said to be *time-invariant* if when $\{y_k\}$ is the output that corresponds to $\{x_k\}$, then $\{y_{k+M}\}$ is the output that corresponds to $\{x_{k+M}\}$.

Suppose that one has a system that is linear and time-invariant. Such systems, often referred to as LTI systems, can be characterized by their response to the delta function:

$$\delta_k \equiv \begin{cases} 1, & k = 0 \\ 0, & \text{otherwise} \end{cases}. \tag{18.2}$$

Suppose that when the input to the system is $\{\delta_k\}$, the output is the sequence $\{h_k\}$. The sequence $\{h_k\}$ is known as the system's *impulse response*.

Suppose that the sequence $\{x_k\}$ is input to an LTI. This sequence can be written

$$\{x_k\} = \sum_{n=-\infty}^{\infty} x_n \{\delta_{k-n}\}$$

where the elements of $\{x_k\}$ are viewed as the weights that multiply the sequences $\{\delta_{k-n}\}$. Making use of linearity and time-invariance, we find that the output sequence that corresponds to this input sequence is

$$\{y_k\} = \sum_{n=-\infty}^{\infty} x_n \{h_{k-n}\}.$$

Clearly, the elements of $\{y_k\}$ are given by

$$y_k = \sum_{n=-\infty}^{\infty} x_n h_{k-n}.$$

The summation *defines* the (non-cyclic) discrete convolution of the elements of two sequences, $\{x_k\}$ and $\{h_k\}$. In what follows, we use an asterisk to denote the convolution operation.

18.5 The Impulse Response and the Transfer Function

Consider the Z-transform of the output of a linear system—of y_k. That is, consider the Z-transform of the discrete convolution of two sequences $\{x_n\}$ and $\{y_n\}$. We find that

$$Y(z) = \sum_{k=-\infty}^{\infty} z^{-k} y_k$$

$$= \sum_{k=-\infty}^{\infty} z^{-k} \sum_{n=-\infty}^{\infty} x_n h_{k-n}$$

$$= \sum_{n=-\infty}^{\infty} x_n \sum_{k=-\infty}^{\infty} z^{-k} h_{k-n}$$

$$= \sum_{n=-\infty}^{\infty} z^{-n} x_n \sum_{k=-\infty}^{\infty} z^{-(k-n)} h_{k-n}$$

$$= X(z)H(z).$$

That is, the Z-transform of the output of an LTI system is the product of the Z-transform of its input and the Z-transform of its impulse response. The Z-transform of the impulse response, $H(z)$, is said to be the system's *transfer function*. The transfer function, $H(z)$, satisfies

$$\frac{Y(z)}{X(z)} = H(z).$$

The ROC of $\{y_k\}$'s Z-transform is at least the intersection of the ROCs of $X(z)$ and $H(z)$. (This last statement has not been proved and is connected to a proper justification of interchanging the order of the summations in the derivation. The justification makes use of Fubini's theorem for sums.)

18.6 A Simple Example

Consider a summer—a system whose output at "time" k, y_k, is the sum of its inputs, x_k, until time k. That is,

$$y_k = \sum_{-\infty}^{k} x_n. \tag{18.3}$$

One way to find this system's transfer function is to find its impulse response and then determine the Z-transform of the impulse response.

If one inputs an impulse—a sequence whose value is 1 when $k = 0$ and is zero otherwise—to the system, the system's output will be 0 until the impulse arrives at $k = 0$, and from $k = 0$ onward the output will be 1. If $x_k = \delta_k$, then $y_k = u_k$. The impulse response is the discrete-time unit step function. As the Z-transform of the unit step is $z/(z - 1)$, we find that the transfer function of the summer is

$$H(z) = \frac{z}{z - 1}$$

and its ROC is $|z| > 1$.

18.7 The Inverse Z-transform

18.7.1 Inversion by Contour Integration

Considering the form of the Z-transform, recalling that [3]

$$\oint_{|z|=R} z^n \, dz = \begin{cases} 0 & n \neq -1 \\ 2\pi j & n = -1 \end{cases},$$

and making use of the uniform convergence of the Z-transform in its ROC, it is easy to see that

$$y_k = \frac{1}{2\pi j} \oint_{|z|=R} z^{k-1} Y(z) \, dz$$

for any circle, $|z| = R$, contained in the interior of the ROC of $Y(z)$.

Though one rarely uses this formula to find the inverse Z-transform, let us consider two interesting examples of its use. First, consider

$$Y(z) = \frac{z}{z - 1}, \qquad |z| > 1.$$

We find that

$$y_k = \frac{1}{2\pi j} \oint_{|z|=R} z^{k-1} \frac{z}{z - 1} \, dz$$

where $R > 1$. As the curve over which we are integrating includes both $z = 0$ and $z = 1$, we must consider the poles at $z = 1$ and $z = 0$. We note that for

$k \geq 0$ there is no pole at $z = 0$; the only pole is at $z = 1$. The residue at this point is 1 and the value of the integral is also one. Thus, $y_k = 1$, $k \geq 0$. For $k < 0$, there is still a residue of 1 at $z = 1$. In addition, however, there is a residue of -1 at $z = 0$. (See Exercise 3 for a proof of this claim.) As the sum of the residues is zero, the value of the integral is zero. That is, $y_k = 0$, $k < 0$. This agrees with the results of Section 18.3.1.

Now, consider the Z-transform

$$Y(z) = \frac{z}{z-1}, \qquad |z| < 1.$$

This time, we are calculating the integral

$$y_k = \frac{1}{2\pi j} \oint_{|z|=R} z^{k-1} \frac{z}{z-1} \, \mathrm{d}z$$

where $R < 1$. In this case, when $k \geq 0$ the integrand is analytic inside the curve, and its integral is zero. For $k < 0$, there is only one residue in the region—the residue at $z = 0$. As the value of this residue is -1, we find that the value of the integral is -1. Thus, for $n < 0$, $y_k = -1$. This agrees with the results of Section 18.3.2.

The existence of the contour integral-based formula guarantees that given a Z-transform and its region of convergence, it is possible to invert the Z-transform. In particular, if two Z-transforms are the same and one's ROC includes the other's, then the two transforms must be the transform of the same sequence. (This is so because the integral that is used in the inverse transform will give the same sequence for any circle in the part of the ROC that is common to both Z-transforms.) In particular, *if two transforms whose functional parts are the same have ROCs that extend to infinity, then the two transforms are actually transforms of the same sequence, and all terms in the sequence that correspond to negative values of the index are zero.*

18.7.2 Inversion by Partial Fractions Expansion

The method that is generally used to invert the Z-transform is to decompose a complicated Z-transform into a sum of less complicated transforms. The less complicated transforms are then inverted by inspection—by making use of our knowledge of simple Z-transforms. We assume that the Z-transform is a rational function of z—that it is a quotient of polynomials in z. Then, we make use of the partial fractions expansion [17] to subdivide the Z-transform into simpler component parts. Finally, we invert the component parts.

Suppose, for example, that

$$Y(z) = \frac{z}{(z-1)(z-1/2)}, \qquad |z| > 1.$$

We can write

$$\frac{Y(z)}{z} = \frac{1}{(z-1)(z-1/2)} = \frac{A}{z-1} + \frac{B}{z-1/2}.$$

(We divide $Y(z)$ by z in order to ensure that when we *finish* separating the rational function into partial fractions, we will be left with a representation for $Y(z)$ in which every fraction has a z in its numerator. These simpler functions will have transforms of the same form as the transforms we examined previously.) Multiplying the fractions by $(z-1)(z-1/2)$, we find that

$$1 = A(z-1/2) + B(z-1).$$

We find that $A + B = 0$ and $A/2 + B = -1$. Thus, $A = 2$ and $B = -2$. We find that

$$Y(z) = \frac{2z}{z-1} - \frac{2z}{z-1/2}, \qquad |z| > 1.$$

As the region of convergence extends to infinity, we know that for all $k < 0$, $y_k = 0$. Also, we know that the inverse of the first part is twice the unit step. Finally, the inverse of the second part is easily seen to be $-2(1/2)^k u_k$. Making use of the results of Section 18.3.3, we find that

$$y_k = 2(1 - (1/2)^k)u_k.$$

18.7.3 Using MATLAB to Help

The MATLAB® command residue can be used to calculate the partial fractions expansion. The format of the command is [R P K] = residue(B,A) where the arrays B and A contain the coefficients of the numerator and the denominator, respectively, and the arrays R, P, and K are the coefficients of the fractions, the poles in each of the fractions, and the coefficients of the polynomial that results from the long division of the original function. K is often the empty array, [].

Let us use MATLAB to calculate the partial fractions expansion of the previous section. With

$$\frac{Y(z)}{z} = \frac{1}{(z-1)(z-1/2)} = \frac{1}{z^2 - 1.5s + 0.5},$$

we find that B = [1] and A = [1 -1.5 0.5]. Giving MATLAB the commands B = [1], A = [1, -1.5, 0.5], and residue(B,A), we find that MATLAB responds with

    R =

        2
       -2

P =

    1.0000
    0.5000

K =

   []

Translating these values back into functions, we find that

$$\frac{Y(z)}{z} = \frac{2}{z - 1.0000} - \frac{2}{z - 0.5000}.$$

This is what we found previously.

18.8 Stability of Discrete-time Systems

A *causal* system is a system whose impulse response, h_k, is zero for all $k < 0$. That is, the impulse response of a causal system does not start until the impulse actually arrives. The ROC corresponding to the transfer function of a causal system extends out to infinity; it includes a region of the form $|z| > R$. All actual systems have to be causal; there is no way that a response to an event can start before the event occurs.

A system is said to be bounded input-bounded output (BIBO) stable if for any bounded input sequence, the output sequence is bounded as well. We would like to determine when a causal system is stable.

The output of a causal LTI system whose impulse response is $\{h_k\}$ is

$$y_k = \sum_{n=0}^{\infty} h_n x_{k-n}.$$

The way to maximize this value for a bounded sequence—a sequence for which $|x_k| \leq C$—is to choose the x_k such that $x_{k-n} h_n$ is non-negative and such that the magnitude of x_{k-n} is as large as possible—in this case, C. For such a bounded input, we find that the output, y_n, is uniformly bounded by

$$\sum_{n=0}^{\infty} C|h_n| \tag{18.4}$$

where each of the x_k is assumed to have magnitude C. If this sum is bounded, then the system is BIBO stable. If not, then it is possible to produce a bounded sequence of inputs [15, Problem 3.21] for which the output is arbitrarily large; thus, the system is not BIBO stable.

Let us assume that the transfer function of our causal LTI system, $H(z)$, is a rational function of polynomials. As the system is causal, the Z-transform of its impulse response is zero for all negative "time," and the ROC that corresponds to it is of the form $|z| > c$. (See Section 18.2.1.) In such a case, the poles of the function determine what kind of sequence $\{h_k\}$ is.

Consider the partial fractions decomposition of $H(z)$. Assuming that all poles of $H(z)$ are simple poles—are poles of order one—we find that

$$H(z) = A_1 \frac{z}{z - \alpha_1} + \cdots + A_N \frac{z}{z - \alpha_N}.$$

Because the system is assumed causal, we know that the ROC includes a region of the form $|z| > C$. Thus, the inverse Z-transform of the transfer function is of the form

$$h_k = (A_1(\alpha_1)^k + \cdots + (\alpha_N)^k)u_k.$$

If there are any poles outside of the unit circle, then it is easy to see that h_k grows without bound. In particular, (18.4) is not bounded, so the system is not BIBO stable. Moreover, suppose that $H(z)$ has poles on the unit circle. Such poles, it is easy to see, correspond to undamped complex exponentials. As the sum of their absolute values is infinite, the system is still not stable.

Finally, if all of the poles are located inside the unit circle, then it is easy to show that h_k decays exponentially quickly, and the sum of the absolute values is bounded. Thus, the system is stable. In sum, we find that *a necessary and sufficient condition for the BIBO stability of a causal LTI system (whose transfer function is a rational function) is that all of the poles of the system lie within the unit circle.* (Though our proof only applies to systems with simple poles, it is easy to extend the proof to the general case.)

18.9 From Transfer Function to Recurrence Relation

When *analyzing* a system, it is generally convenient to work with transfer functions. When *implementing* a system, one generally needs a recurrence relation—one that gives the next value of the output of a system in terms of the current and previous values of the input and the previous values of the output. It is simple to go from one form to the other.

Let $X(z)$ be the Z-transform of the input sequence, let $Y(z)$ be the Z-transform of the output sequence, let $H(z)$ be the transfer function of a causal system, and let us assume that $H(z)$ is a rational function of z. Then, we know that

$$\frac{Y(z)}{X(z)} = H(z) = \frac{a_0 + \cdots + a_n z^n}{b_0 + \cdots + b_m z^m}.$$

Divide both the numerator and the denominator of the transfer function by the highest power of z appearing in the transfer function. Generally speaking, this will be z^m, as transfer functions are generally proper[1]. This will cause the transfer function to be expressed in terms of negative powers of z. We find that

$$\frac{Y(z)}{X(z)} = \frac{a_0 z^{-m} + \cdots + a_n z^{n-m}}{b_0 z^{-m} + \cdots + b_m}.$$

Cross-multiplying, we find that

$$(b_0 z^{-m} + \cdots + b_m)Y(z) = (a_0 z^{-m} + \cdots + a_n z^{n-m})X(z).$$

Making use of the properties of the Z-transform, and inverting the Z-transform, leads to the equation

$$b_0 y_{k-m} + \cdots + b_m y_k = a_0 x_{k-m} + \cdots + a_n x_{k+n-m}.$$

Finally some algebraic manipulation leaves us with

$$y_k = (-b_0 y_{k-m} + \cdots b_{m-1} y_{k-1} + a_0 x_{k-m} + \cdots + a_n x_{k+n-m})/b_m.$$

In a causal system, the current value of the output must be a function of the current and previous value of the input and the previous values of the output—and that is just what we have.

When implementing a digital filter using a microprocessor, one generally "declares" that the first sample processed by the microprocessor arrived at time—at index—zero. The recurrence relation for y_0 requires the current value of the input and *previous values* of the input and the output. Clearly the previous values of the input should be taken to be zero. In a linear, causal system, the output of the system before any signal is input to the system must be zero. We see that the initial conditions of the system are all zero. That is,

$$x_k = y_k = 0, \qquad k = -1, \ldots, -m.$$

Let us consider a simple example of how one converts a transfer function into a recurrence relation. Given the transfer function of the summer of Section 18.6,

$$H(z) = \frac{z}{z-1},$$

let us find the recurrence relation satisfied by the input to, and the output of, the summer. We know that

$$\frac{Y(z)}{X(z)} = \frac{z}{z-1} = \frac{1/z}{1/z}\frac{z}{z-1} = \frac{1}{1-z^{-1}}.$$

Cross-multiplying, we find that

$$(1 - z^{-1})Y(z) = X(z).$$

[1] A transfer function is said to be *proper* if it is a rational function and the degree of its denominator is greater than, or equal to, the degree of its numerator.

Converting back to an equation in the time domain, we find that

$$y_k - y_{k-1} = x_k \Rightarrow y_k = x_k + y_{k-1}.$$

Considering (18.3) and noting that

$$y_{k-1} = \sum_{n=-\infty}^{k-1} x_n,$$

we find that our recurrence relation is indeed satisfied by the output of the summer.

18.10 The Sinusoidal Steady-state Response of Discrete-time Systems

Consider a stable LTI system whose transfer function is $H(z)$. Suppose that the system accepts samples that are taken every T seconds. If the input to the system is a complex exponential with angular frequency $\omega = 2\pi f$, the samples seen by the system are

$$x_k = \begin{cases} e^{j\omega kT} & k \geq 0 \\ 0 & k < 0 \end{cases}.$$

As the non-zero terms are of the form α^k for $\alpha = e^{j\omega T}$, the Z-transform of this sequence is simply

$$X(z) = \frac{z}{z - e^{j\omega T}}.$$

The Z-transform of the output of the system, $Y(z)$, is

$$Y(z) = H(z)\frac{z}{z - e^{j\omega T}}.$$

Consider the partial fractions expansion of the output. It contains two sets of terms. One term has the same pole as the input. All the rest "inherit" the poles of $H(z)$—all of which lie inside the unit circle.

Consider the inverse Z-transform of $Y(z)$. All the terms whose poles are inherited from the transfer function correspond to damped exponentials. Thus, in the steady state they make *no* contribution. In order to determine the steady-state behavior, we need only know the coefficient of

$$\frac{z}{z - e^{j\omega T}}.$$

Let us calculate this coefficient.
    We know that

$$Y(z) = H(z)\frac{z}{z - e^{j\omega T}} = \frac{Az}{z - e^{j\omega T}} + \text{poles inside unit circle}.$$

Multiplying through by $z - e^{j\omega T}$ and substituting $z = e^{j\omega T}$, we find that

$$A = H(e^{j\omega T}).$$

The sinusoidal steady-state response—the response after all the transients have died down—is $H(e^{j\omega T})$ times the input complex sinusoid.

If the input to the system is

$$x_k = \begin{cases} \cos(\omega kT) & k \geq 0 \\ 0 & k < 0, \end{cases}$$

then by using Euler's formula and the linearity of the Z-transform, we find that the steady-state response is

$$\frac{1}{2} \left[H(e^{j\omega T})e^{j\omega kT} + H(e^{-j\omega T})e^{-j\omega kT} \right].$$

As long as the polynomials that define $H(z)$ have real coefficients, it is easy to show that $\overline{H(z)} = H(\bar{z})$. Thus, we find that the steady-state output is

$$\begin{aligned} \text{steady-state output} &= \frac{1}{2} \left(H(e^{j\omega T})e^{j\omega kT} + H(e^{-j\omega T})e^{-j\omega kT} \right) \\ &= \frac{1}{2} \left(H(e^{j\omega T})e^{j\omega kT} + \overline{H(e^{j\omega T})e^{j\omega kT}} \right) \\ &= \text{Re}(H(e^{j\omega T})e^{j\omega kT}) \\ &= \text{Re}(H(e^{j\omega T}))\cos(\omega kT) - \text{Im}(H(e^{j\omega T}))\sin(\omega T). \end{aligned}$$

This is easily seen to be equal to

$$|H(e^{j\omega T})| \cos\{\omega kT + \angle[H(e^{j\omega T})]\}.$$

That is, the system filters the input sinusoid. It amplifies it in a way that is dependent on ω and it shifts its phase in a way that is dependent on ω.

The function $H(e^{j\omega T})$ is called the *frequency response* of the system and $|H(e^{j\omega T})|$ is known as the system's *magnitude response*. Note that the frequency (and magnitude) response is a periodic function of ω whose period is $2\pi/T$. (This is perfectly reasonable, as sampling a signal aliases frequencies above $f = 1/(2T)$—or $\omega = \pi/T$—to frequencies below $f = 1/(2T)$.) Let us take the region $\omega \in (-\pi/T, \pi/T]$ as the period of interest. Assuming that the polynomials of which $H(z)$ is composed have real coefficients, we know that $H(e^{-j\omega T}) = \overline{H(e^{j\omega T})}$. Thus, the negative frequencies do not provide any new information. In order to understand the entire frequency response, it is sufficient to know the frequency response in the region $\omega \in [0, \pi/T]$. Generally speaking, when a frequency (or magnitude) response is plotted, it is only plotted in this range. (The angular frequency π/T corresponds to the frequency $1/(2T)$—the Nyquist frequency of the system.)

18.11 MATLAB and Linear Time-invariant Systems

MATLAB has two "levels" of commands for helping users analyze and design linear time-invariant systems. One set of commands are standard command-line functions that allow the user to define an LTI system and perform different analyses on LTI systems. The second "class" of functions is a single command that opens a graphical user interface that allows the user to perform many analyses from within a single window.

18.11.1 Individual Commands

Before one can analyze an LTI using MATLAB, one must define the LTI for MATLAB. MATLAB has a command, tf, for defining a transfer function object. Assuming that one would like to describe the discrete-time system

$$H(z) = \frac{a_N z^N + \cdots + a_0}{b_M z^M + \cdots + b_0}$$

whose sampling time is T_s, one gives MATLAB the command

$$H = \text{tf}([a_N \; a_{N-1} \; \cdots \; a_0], [b_M \; b_{M-1} \cdots \; b_0], T_s).$$

If one does not wish to specify a sampling time, then in place of T_s one enters -1.

Suppose, for example, that one wanted to characterize a summer to MATLAB, and suppose that the sampling period of the system is 1 ms. Assuming that one wants to refer to the transfer function as H, one would give MATLAB the command H = tf([1 0],[1 -1], 0.001). MATLAB replies to this command with

```
Transfer function:
  z
-----
z - 1


Sampling time: 0.001
```

Having defined the system to MATLAB, it is now simple to have MATLAB provide all sorts of information about the system. Perhaps the most commonly used function is **bode**. This command causes MATLAB to produce the Bode plots[2] that correspond to a system. As we have mentioned, it is common to plot the frequency response up to the angular frequency π/T.

---

[2] The Bode plots that correspond to a system are plots of the system's magnitude response and the system's phase response. The magnitude response is generally given in dB, and the frequency axis is logarithmic. The phase response is given in degrees, and the frequency axis is logarithmic. The plots are generally given one above the other, with the frequency axes scaled identically.

To display the frequency response of our system, one gives MATLAB the command `bode(H)` (after defining H by using the `tf` command). MATLAB responds with Figure 18.1. The plot ends shortly after $\omega = 3{,}000$. As $T_s = 0.001$, the maximum angular frequency should be $\pi/0.001 \approx 3{,}000$, and the plot's ending immediately past $3{,}000\,\mathrm{rad\,s^{-1}}$ is reasonable. Also, note that as the frequency becomes progressively lower, the magnitude becomes progressively larger. This is also to be expected. Recall that $z = e^0 = 1$ is associated with a constant input. The gain at that point is $H(1) = \infty$.

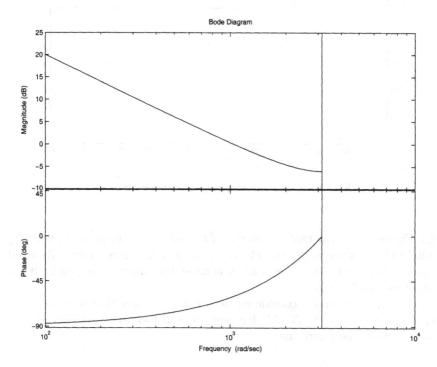

Fig. 18.1. The Bode plots corresponding to the summer

If all that one is interested in is the magnitude response of a system—if the phase response is not important—then one can make use of the `bodemag` command. If one would like to know how a system responds to an impulse, one can give MATLAB the command `impulse`. This command causes MATLAB to plot the impulse response of the system. In our case, giving MATLAB the command `impulse(H)` causes MATLAB to respond with Figure 18.2. As the impulse response of a summer is a unit step, the plot is precisely what we should have expected.

Finally, by giving MATLAB the command `step(H)`, one causes MATLAB to respond with the response of the system to a unit step function. In our case, giving MATLAB the command `step(H)` causes MATLAB to respond

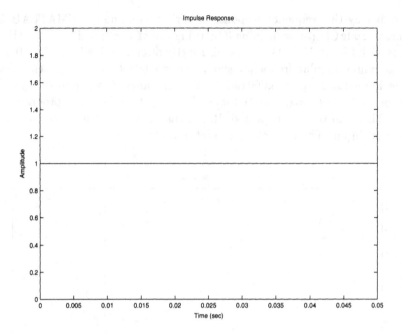

Fig. 18.2. The impulse response of the summer

with Figure 18.3. Note that the output of the summer is an (almost perfectly) straight line—as one would expect from a summer that sees a one at its input from $t = 0$ onward. The slope is 1,000 because the summer samples its input 1,000 times each second.

All of the commands considered have many features that have not even been hinted at here. MATLAB has several help facilities, and they can be used to widen one's horizons.

18.11.2 The ltiview Command

In addition to the commands **bode**, **bodemag**, **impulse**, and **step**, MATLAB has a command called ltiview. This command allows the user to produce and organize many different plots. Typing ltiview with no arguments at the MATLAB prompt causes MATLAB to open a new window. By using the window's drop-down menus, one can request and organize the plots one would like MATLAB to produce.

18.12 Exercises

1. Show that if the ROC of $A(z)$ includes $|z| = 1$, then

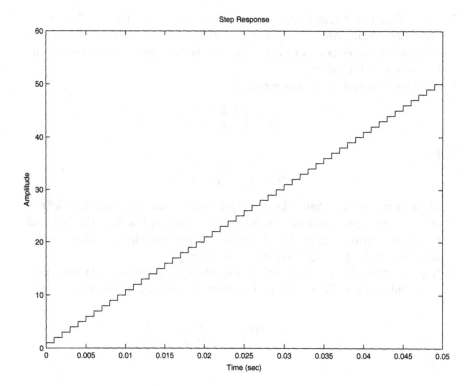

Fig. 18.3. The step response of the summer

$$A(1) = \sum_{-\infty}^{\infty} a_k.$$

2. Show that if

$$\sum_{k=-\infty}^{-1} a_k z^{-k}$$

converges for $z = b$, then it converges for all z that satisfy $|z| < |b|$.

3. Show that the residue of the function

$$\frac{z^{-n}}{z - 1}$$

at $z = 0$ is -1 if $n \geq 1$. (You may find the partial fractions expansion helpful.)

4. Find the Z-transform of the sequence

$$a_k = \begin{cases} 0 & k < 0 \\ \cos(\omega k) & k \geq 0 \end{cases}.$$

5. Prove that the Z-transform is linear by making use of the definition of the Z-transform. Also, show that the ROC of the resulting Z-transform contains the intersection of the ROCs of the two transforms whose combination is being taken.

6. Find the Z-transform of the sequence

$$a_k = \begin{cases} 0 & k < 0 \\ k & k \geq 0 \end{cases}.$$

7. Let

$$a_k = \begin{cases} 1/k & k \geq 1 \\ 0 & k < 1 \end{cases}.$$

Show that the sum that defines $A(1)$ diverges while the sum that defines $A(-1)$ converges conditionally. Show that this implies that the ROC of the Z-transform contains the region $|z| > 1$, and explain in what sense the behavior of $A(z)$ is "odd" on the circle $|z| = 1$.

8. Suppose that the functions below are taken to be the transfer functions of causal systems. Which of the functions correspond to stable systems?

a)

$$H(z) = \frac{z}{z^2 + 1}.$$

b)

$$H(z) = \frac{z}{z + 1/2}.$$

c)

$$H(z) = \frac{z}{z - 1/2}.$$

d)

$$H(z) = \frac{z}{z^2 - z/2 + 1/4}.$$

9. Find the sequence that corresponds to each of the following Z-transforms

a)

$$H(z) = \frac{z}{(z - 1/2)(z + 1/2)}, \qquad |z| > 1/2.$$

b)

$$H(z) = \frac{z}{(z - 1/2)^2}, \qquad |z| > 1/2.$$

c)

$$H(z) = \frac{z}{(z - 1/2)^2}, \qquad |z| < 1/2.$$

10. Find the Z-transforms of the following sequences
    a)

$$\{a_k\} = \{0, 0, \ldots, 0, 1, \ldots, k^2, \ldots\} = \{k^2 u_k\}.$$

b)

$$\{a_k\} = \{\ldots, 1/2, 1, 1/2, \ldots, \} = \left\{\frac{1}{2^{|k|}}\right\}.$$

11. Sketch the magnitude response of the filter whose transfer function is

$$H(z) = \frac{z}{z - 1/2}.$$

   Check your answer using the MATLAB **bodemag** command. (You will need to set $T_s = -1$ in the **tf** command.)

12. Calculate the Z-transform of the sequence $\{\delta_k\}$.

13. Find the recurrence relation that corresponds to the causal systems described by each of the following transfer functions. Let the sequence input to the system be x_k and the sequence output by the system be y_k.
    a)

$$H(z) = \frac{z}{z^2 - 1}.$$

b)

$$H(z) = \frac{z - 1/2}{z^2 - z/2 - 1/2}.$$

14. Calculate the inverse Z-transform of

$$A(z) = \frac{z}{z - \alpha}, \qquad |z| < \alpha.$$

You may wish to write the function as a Taylor series in power of z/α.

19

Filter Types

Summary. Analog filters are generally built using resistors, capacitors, inductors, and operational amplifiers. The impulse response of such filters begins when the impulse arrives and continues indefinitely. A filter with an impulse response of this type is known as an infinite impulse response (IIR) filter.

In this chapter, we discuss the filter types available to the engineer designing digital filters. Some digital filters are IIR filters, while others are finite impulse response (FIR) filters. An FIR filter is a filter whose impulse response is identically zero after a certain time.

Keywords. impulse response, IIR, FIR.

19.1 Finite Impulse Response Filters

Suppose that the impulse response of a digital filter is

$$h_n, \qquad |n| \leq N,$$

and is zero for all n outside of this range. Filters of this sort—filters whose impulse response consists of a finite number of terms—are called finite impulse response (FIR) filters. By making use of the principle of superposition, it is easy to see that the response, y_n, of the filter to a generic signal x_n is

$$y_n = \sum_{k=-\infty}^{\infty} x_k h_{n-k}$$

$$= x_{n-N} h_N + x_{n-(N-1)} h_{N-1} + \cdots + x_n h_0 + \cdots + h_{-N} x_{n+N}.$$

That is, y_n is a weighted sum of samples of the input. The transfer function of an FIR filter can be expressed as

$$H(z) = h_{-N} z^N + \cdots + h_{-1} z^1 + h_0 + h_1 z^{-1} + \cdots + h_N z^{-N}$$

$$= \frac{z^{2N} h_{-N} + \cdots + z^N h_0 + \cdots + h_N}{z^N}, \qquad 0 < |z| < \infty.$$

For any bounded input sequence $\{x_n\}$, for any input sequence for which $|x_n| \leq C$, we find that

$$|y_n| \leq |h_{-N}x_{n+N} + \cdots + h_0 + \cdots h_N x_{n-N}|$$
$$\leq (|h_{-N}| + \cdots + |h_N|) C.$$

Thus, the output of the FIR filter is bounded whenever the input is. We find that *FIR filters are always stable.*

19.2 Infinite Impulse Response Filters

Infinite impulse response (IIR) filters are filters whose impulse response persists indefinitely. They are precisely those filters whose transfer functions *do* contain finite non-zero poles. For example, the function

$$H(z) = \frac{1}{z - 1/2}, \qquad |z| > 1/2$$

is the transfer function of an IIR filter. Its impulse response is easy to calculate. It is easy to see that (for $|z| > 1/2$) we have

$$H(z) = \frac{1}{z - 1/2}$$
$$= z^{-1} \frac{1}{1 - z^{-1}/2}$$
$$= z^{-1} \sum_{k=0}^{\infty} (z^{-1}/2)^k$$
$$= (1/2)^0 z^{-1} + (1/2)^1 z^{-2} + \cdots + (1/2)^{k-1} z^{-k} + \cdots.$$

Thus, the impulse response of the filter is

$$h_k = \begin{cases} 0 & k \leq 0 \\ (1/2)^{k-1} & k \geq 1 \end{cases}.$$

19.3 Exercises

1. a) What is the impulse response of the filter whose transfer function is

$$H(z) = \frac{T_s z}{z - 1}, \qquad |z| > 1?$$

(Here T_s is the sampling period.)

b) Is the filter whose transfer function is given by $H(z)$ an FIR filter or an IIR filter?

c) Find the filter's output, y_k, as a function of its input, x_k. You may assume that $x_k = 0$ for all $k < 0$.

d) What operation does this filter approximate when $T_s << 1$?

2. What is the impulse response of the filter defined by

$$H(z) = \frac{z}{z + 1/2}, \qquad |z| > 1/2?$$

Is this filter an FIR or an IIR filter? Explain!

3. What is the impulse response of the filter defined by

$$H(z) = \frac{z}{z + 1/2}, \qquad |z| < 1/2?$$

Is this filter an FIR or an IIR filter? Explain!

When to Use C (Rather than Assembly Language)

Summary. In many introductory microprocessor (or microcontroller) courses, one is taught to program in assembly language. Until now, it was probably best to program in assembly language, but at this point it is probably best to move to C. In this chapter, we discuss why one might want to make this change, and we give a first example of implementing a filter using a program written in C.

Keywords. assembly language, C, low-pass filter, RC filter.

20.1 Introduction

In introductory courses in microprocessors, one often spends much of one's time controlling the processor and its peripherals. Assembly language is built for such tasks and makes them relatively easy. In particular, the machine language of the 8051 family of microprocessors supports many instructions that work on individual bits. When every bit is significant—as it is when one is setting up an SFR—this ability is very important.

In several of the early laboratories, we used a microprocessor as a data-gathering tool. We set up its SFRs and let the microprocessor run. In such a case, it makes sense to use assembly language. In the laboratories that follow, we use the microprocessor (often a microcontroller) to do calculations. When one is interested in relatively heavy-duty calculations, it often makes sense to move to a high-level language. In our case, that language is C. For more information about the version of C we used with the ADuC841, see [13].

20.2 A Simple Low-pass Filter

In order to make the first C program relatively simple, we consider a simple infinite impulse response (IIR) filter. Perhaps the simplest IIR digital filter is the (causal) filter defined by the equation

$$y_n = (1 - \alpha)y_{n-1} + \alpha x_n, \qquad 0 < \alpha < 1, \quad y_{-1} = 0.$$

The Z-transform of $\{y_n\}$, $Y(z)$, satisfies the equation

$$Y(z) = (1 - \alpha)z^{-1}Y(z) + \alpha X(z).$$

We find that

$$Y(z) = \frac{\alpha z}{z - (1 - \alpha)} X(z),$$

and the filter's transfer function is

$$T(z) = \frac{\alpha z}{z - (1 - \alpha)}.$$

Clearly, $T(1) = 1$, and the (only) pole of this system is $(1 - \alpha)$. To keep the pole inside the unit circle—to keep the filter stable—we find that $|(1 - \alpha)| < 1$ or that $0 < \alpha < 2$.

20.3 A Comparison with an *RC* Filter

The transfer function of an *RC* low-pass filter is

$$T(s) = \frac{1}{RCs + 1}.$$

If $y(t)$ is the output of the filter and $x(t)$ is the filter's input, then (assuming that $y(0) = 0$) we find that [6]

$$RCy'(t) + y(t) = x(t).$$

If we want an analogous discrete-time system, then it is reasonable to consider samples of the input, $x_n = x(nT_s)$, and to let the output, y_n, satisfy

$$RC\frac{y_n - y_{n-1}}{T_s} + y_n = x_n. \tag{20.1}$$

The term

$$\frac{y_n - y_{n-1}}{T_s}$$

should tend to y' as $T_s \to 0$. Rearranging terms in (20.1), we find that

$$y_n = \frac{RC}{RC + T_s}y_{n-1} + \frac{T_s}{RC + T_s}x_n, \qquad y_0 = 0. \tag{20.2}$$

If we let $\alpha = \frac{T_s}{RC + T_s}$, then this is precisely the type of filter we have been dealing with. Note that in (20.2) we find that $0 < \alpha < 1$, whereas the filter is actually stable for $0 < \alpha < 2$.

20.4 The Experiment

Write a C program to implement the filter of Section 20.2. It should be possible to achieve a sample rate of 5,000 samples s^{-1}. Try a variety of values of α. Make sure that some of the values exceed one. For a square wave input, what does the output look like when $\alpha < 1$? What about when $\alpha > 1$? In what way(s) is the output when $\alpha > 1$ similar to the output of a standard RC low-pass filter? In what way(s) is it different?

Compare the C code with the computer-generated assembly language code. (To cause the computer to include the assembly code in the listing when using the Keil μVision3 IDE, you must change the *options* for the project.) Note that in the sections where one is setting up the registers, the assembly language code is not much longer than the C code. However, in the regions where the C code is performing calculations, the assembly language code is *much* longer.

20.5 Exercises

1. Find and plot the magnitude response of the filter described in Section 20.2 when $\alpha = 3/2$.
2. Find the impulse response of the causal filter whose transfer function is

$$T(z) = \frac{\alpha z}{z - (1 - \alpha)}, \qquad 0 < \alpha < 2.$$

   What qualitative change does the impulse response undergo when α goes from being less than one to being more than one?
3. Show that for $1 < \alpha < 2$, the causal filter whose transfer function is

$$T(z) = \frac{\alpha z}{z - (1 - \alpha)}$$

   is a high-pass filter.
4. a) Use Simulink® to simulate the "RC-type" filter of Sections 20.2 and 20.3 for a variety of values of $0 < \alpha < 1$ and $1 < \alpha < 2$. Let the sampling period be 0.01 s.
   b) Examine the filter's response to a variety of low- and high-frequency sinewaves both when the system includes an anti-aliasing filter and when it does not.
   c) Use the MATLAB command **bode** to examine the frequency response of the filters that were implemented. To what extent do the simulation results agree with the information presented on the Bode plots?

Two Simple FIR Filters

Summary. Having seen (and implemented) a simple IIR filter, we consider two simple FIR filters. In this chapter, we develop both a simple FIR low-pass filter and a simple FIR high-pass filter.

Keywords. averaging filter, high-pass filter, low-pass filter, FIR filters.

21.1 Introduction

The averaging filter is a simple finite impulse response (FIR) low-pass filter. Let F_s be the sampling rate of the system, and let T_s be its reciprocal, the sampling period. Let $x_k = x(kT_s)$ be the kth sample of the input (where k starts from 0), and let y_k be the output of the filter at time $t = kT_s$. Then, an N-coefficient (also known as an N-*tap*) low-pass filter is defined by the equation

$$y_n = \frac{x_n + x_{n-1} + \cdots + x_{n-N+1}}{N}.$$

This filter's output, y_n, is the running average of the last N samples of the input, x_n.

The Z-transform of the sequence $\{y_n\}$ is

$$Y(z) = (1 + z^{-1} + \cdots + z^{-N+1})X(z)/N. \tag{21.1}$$

Summing the finite geometric series, we find that the transfer function of the filter, $T_{\mathrm{LP}}(z)$, is

$$T_{\mathrm{LP}}(z) = \frac{1}{N}\frac{1 - z^{-N}}{1 - z^{-1}}.$$

Substituting $z = e^{2\pi j f T_s}$, we find that the frequency response of the filter is

$$T_{\mathrm{LP}}(e^{2\pi jfT_s}) = \frac{1}{N}\frac{1 - e^{-2N\pi jfT_s}}{1 - e^{-2\pi jfT_s}}$$

$$= \frac{1}{N}e^{-(N-1)\pi jfT_s}\frac{\sin(N\pi fT_s)}{\sin(\pi fT_s)}.$$

Making use of L'hôpital's rule (or, more properly, not using the formula for the geometric series when $f = k/T_s$ but simply summing the series), we find that at $f = 0$ the transfer function is 1; at no other point is it greater. (See Exercise 3.)

The simplest FIR high-pass filter is given by the equation

$$y_n = \frac{1}{N}\left(x_n - x_{n-1} + \cdots + (-1)^k x_{n-k} + \cdots - x_{n-N+1}\right)$$

where N is even. Passing to the Z-transform of $\{y_n\}$, $Y(z)$, we find that

$$Y(z) = \frac{1}{N}\left(1 - z^{-1} + \cdots + (-1)^k z^{-k} + \cdots - z^{-N+1}\right)X(z).$$

The transfer function of the filter, $T_{\mathrm{HP}}(z)$, is

$$T_{\mathrm{HP}}(z) = \frac{1}{N}\left(1 - z^{-1} + \cdots + (-1)^k z^{-k} + \cdots - z^{-N+1}\right)$$

$$= \frac{1}{N}\frac{1 - (-1)^N z^{-N}}{1 + z^{-1}}. \tag{21.2}$$

As by assumption N is even, we find that

$$T_{\mathrm{HP}}(z) = \frac{1}{N}\frac{1 - z^{-N}}{1 + z^{-1}}.$$

Substituting $z = e^{2\pi jfT_s}$, we find that the frequency response of the filter is

$$T_{\mathrm{HP}}(e^{2\pi jfT_s}) = \frac{1}{N}\frac{1 - e^{-2N\pi jfT_s}}{1 + e^{-2\pi jfT_s}}$$

$$= \frac{1}{N}\frac{e^{-N\pi jfT_s}}{e^{-\pi jfT_s}}\frac{e^{N\pi jfT_s} - e^{-N\pi jfT_s}}{e^{\pi jfT_s} + e^{-\pi jfT_s}}$$

$$= \frac{1}{N}e^{-(N-1)\pi jfT_s}j\frac{\sin(N\pi fT_s)}{\cos(\pi fT_s)}.$$

Note that when $f = 0$, the frequency response of this filter is zero. The filter does not pass DC at all. Plugging $z = e^{2\pi j(1/(2T_s))T_s} = -1$ into (21.2), it is easy to see that when $f = 1/(2T_s)$—which is the "Nyquist frequency" for the filter—the filter's frequency response is 1. In the sinusoidal steady state, the filter passes high-frequency signals without altering them.

21.2 The Experiment

Please modify the program that implements a simple IIR filter to implement the simple averaging filter with five elements in the average. A sample rate of 5,000 samples s^{-1} should be achievable. Note that there is no need to buffer the samples of the output—they are never reused. (That is an *essential* difference between FIR and IIR filters.)

Please examine the frequency response of the filter. Use a sinewave generator and an oscilloscope. Does practice agree with theory?

Having finished with the averaging filter, implement the simple high-pass filter with $N = 6$. Once again, a sample rate of 5,000 samples s^{-1} should be achievable. Using a sinewave generator and an oscilloscope, examine the filter's frequency response. Do practice and theory agree?

21.3 Exercises

1. Consider a simple averaging filter with five taps—when $N = 5$.
   a) Plot the magnitude response of the filter as a function of fT_s.
   b) At what frequencies do the zeros of the transfer function occur?
   c) Give an intuitive explanation of the answer to the previous section.
2. Consider the six-tap high-pass filter—the filter for which $N = 6$.
   a) Plot the magnitude response of the filter as a function of fT_s.
   b) At what frequencies do the zeros of the transfer function occur?
   c) Give an intuitive explanation of the answer to the previous section for at least one of the zeros of the transfer function.
3. Make use of (21.1), the triangle inequality, and the definition of the frequency response to show that

$$|T_{LP}(e^{2\pi jfT_s})| \le 1.$$

Very-narrow-band Filters

Summary. In the preceding chapter, two simple FIR filters were presented. In this chapter, we consider an IIR bandpass filter. In the design of the IIR bandpass filter, a simple FIR notch filter is used. By adding feedback to the system, a programmable-bandwidth IIR filter is produced.

Keywords. notch filter, programmable-bandwidth filter, feedback.

22.1 A Very Simple Notch Filter

Consider the FIR filter whose transfer function is

$$T_{\text{simple}}(z) = (1 - e^{2\pi jFT_s}z^{-1})(1 - e^{-2\pi jFT_s}z^{-1}) = 1 - 2z^{-1}\cos(2\pi FT_s) + z^{-2}.$$

It is clear that the frequency response of this filter has exactly two zeros, and they are located at $z = e^{\pm 2\pi jFT_s}$. That is, this filter has a "notch"—a zero in its frequency response—at $\pm F$ Hz.

22.2 From Simple Notch to Effective Bandpass

Consider the system of Figure 22.1. Clearly,

$$V_{\text{out}}(z) = V_{\text{in}}(z) - KT_{\text{simple}}(z)V_{\text{out}}(z).$$

Manipulating this equation allows us to show that the transfer function of the system with feedback is

$$T_{\text{feedback}}(z) = \frac{V_{\text{out}}(z)}{V_{\text{in}}(z)} = \frac{1}{1 + KT_{\text{simple}}(z)}.$$

$T_{\text{simple}}(z)$ has only the two zeros found above. Thus, when $f = \pm F$ Hz, the frequency response is 1. If K is sufficiently large, it is easy to see that at all frequencies far enough from $\pm F$, the transfer function will be approximately 0. That is, this filter blocks all frequencies except those very near $\pm F$—this is a narrow-band bandpass filter. By making K sufficiently large, the filter can be made as narrow-band as desired.

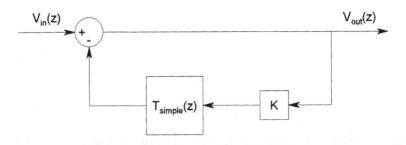

Fig. 22.1. An interesting use of feedback

22.3 The Transfer Function

It is easy to see that

$$T_{\text{feedback}}(z) = \frac{1}{1 + K(1 - 2\cos(2\pi F T_s)z^{-1} + z^{-2})}.$$

This leads to a simple formula for y_k (the current value of the output) in terms of x_k (the current value of the input) and the previous values of the output. (See Exercise 2.)

22.4 The Experiment

Use the technique presented in this chapter to implement a bandpass filter that passes $1{,}350\,\text{Hz}$. Let the sampling rate of the filter be $5{,}400\,\text{samples s}^{-1}$. Examine the filter with a variety of gain values. Use a digital oscilloscope to store the output of the filter at $1{,}450\,\text{Hz}$ for each of the gains you implement. Label the output, and submit it with the program.

22.5 Exercises

1. Explain why putting $KT_{\text{simple}}(z)$ in the "forward path" (in the upper portion of the system), rather than in the feedback loop, gives a notch filter that removes F Hz signals. (Give a reasonably intuitive explanation.)
2. Find the equation that expresses the output of the filter described in this chapter in terms of the filter's previous outputs and the filter's current and previous inputs.
3. Show that the filter of Section 22.3 is stable for all $K > 0$.
4. Examine the magnitude response of the effective bandpass filter when $K = 10$, $F = 1\,\text{kHz}$, and $T_{\text{s}} = 1/5{,}400\,\text{s}$. Note that the maximum value of the magnitude is not $0\,\text{dB}$, and it does not occur at precisely $1\,\text{kHz}$. Explain how this can happen and why this does not contradict the theory developed in this chapter.

Design of IIR Digital Filters: The Old-fashioned Way

Summary. In this chapter, we discuss a systematic method of designing low-pass infinite impulse response (IIR) digital filters. As this procedure requires that one first design an analog low-pass filter, we start by discussing the properties of analog low-pass filters, and we proceed to techniques for designing such filters. Then, we consider one technique for converting such filters into digital low-pass filters.

Keywords. analog filters, Butterworth filter, Chebyshev filter, elliptic filter, bilinear transform.

23.1 Analog Filter Design

There are three very standard types of analog filters: Butterworth filters, Chebyshev filters, and elliptic filters. As this is not a course in analog filter design, we only describe the filters briefly.

Butterworth filters are also called "maximally flat" filters. The square of the magnitude of a low-pass Butterworth filter satisfies the equation

$$|H_B(j\omega)|^2 = \frac{1}{1 + (j\omega/j\omega_c)^{2N}}.$$

By making use of the fact that for small $|x|$ we know that $1/(1+x) \approx 1 - x$, we find that near $\omega = 0$ we can approximate the square of the magnitude by

$$|H_B(j\omega)|^2 \approx 1 - (\omega/\omega_c)^{2N}.$$

For values of ω that are much smaller than ω_c, this is *very* close to one. The filter is *very* flat near $\omega = 0$.

Chebyshev filters have the form

$$|H_C(j\omega)|^2 = \frac{1}{1 + \epsilon^2 V_N^2(\omega/\omega_c)}$$

where $V_N(x)$ are the Chebyshev polynomials, and ω_c is the cut-off frequency of the filter[1]. For $|x| \leq 1$, the Chebyshev polynomials are defined by the equation

$$V_N(x) = \cos(N \cos^{-1}(x)).$$

It is easy to see that $V_N(x)$ is a polynomial when $1 \geq x \geq -1$. For example, with $N = 1$, we have $V_1(x) = x$, and with $N = 2$, we have

$$V_2(x) = \cos(2 \cos^{-1}(x)) = 2 \cos^2(\cos^{-1}(x)) - 1 = 2x^2 - 1.$$

In Exercise 4, it is shown that

$$\cos((N+1)\theta) = 2\cos(\theta)\cos(N\theta) - \cos((N-1)\theta).$$

Making use of this identity, we find that

$$
\begin{aligned}
V_{N+1}(x) &= \cos((N+1)\cos^{-1}(x)) \\
&= 2\cos(\cos^{-1}(x))\cos(N\cos^{-1}(x)) - \cos((N-1)\cos^{-1}(x)) \\
&= 2xV_N(x) - V_{N-1}(x).
\end{aligned}
$$

By a simple induction, we see that for all $N \geq 1$, the functions $V_N(x)$ are Nth-order polynomials. Though we derive the polynomials by considering $|x| \leq 1$, the polynomials we find are, of course, defined for all x.

By noting that $N \cos^{-1}(x)$ goes from 0 to $N\pi$ as x goes from 1 to -1, we find that $\cos(N \cos^{-1}(x))$ passes through zero N times in this region. Thus, all N zeros of the Nth-order polynomial $V_N(x)$ are located between -1 and 1. This shows that outside this region the polynomial cannot ever return to zero. Though we have not proved it, outside $[-1, 1]$ the polynomial $V_N^2(x)$ increases monotonically. (See Exercise 5 for a proof of this fact.)

We see that $V_N(\omega/\omega_c)$ has ripples—oscillates—in the passband but not the stopband. So does the filter designed using the Chebyshev polynomials. Because we allow some ripples, we are able to make the transition region narrower for the same-order filter. (There is a second type of Chebyshev filter with ripples in the stopband but not in the passband.) See Figure 23.1 for a comparison of the magnitude responses of several filters of different types and orders.

The design of elliptic filters is more complicated, and we do not consider it here. We do, however, describe the most important properties of elliptic filters. Elliptic filters have ripples in the passband and the stopband. In that sense, they are not very ideal filters. On the other hand, they have the narrowest transition region of the three filter types presented. If keeping the filter order small is important, then elliptic filters are often the way to go.

---

[1] For Chebyshev filters, the cut-off frequency is *not* generally the 3 dB down point— the point at which the magnitude of the filter's response falls to $1/\sqrt{2}$ of its maximum value.

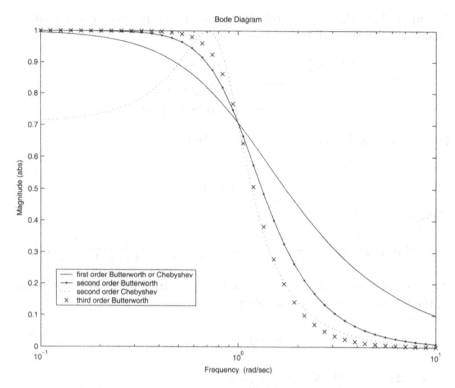

Fig. 23.1. The magnitude responses corresponding to several different filters. The common cut-off frequency of the filters is $1\,\mathrm{rad\,s^{-1}}$. In the case of the Chebyshev filters, $\epsilon = 1$.

23.2 Two Simple Design Examples

Let us consider the Butterworth filters that correspond to $N = 1$ and $N = 2$. Until now, we have been considering $|H_{\mathrm{B}}(j\omega)|^2$. By definition, $H(j\omega) = H(s)|_{s=j\omega}$. Also, note that given a rational function with real coefficients

$$H(s) = \frac{a_0 + \cdots + a_n s^n}{b_0 + \cdots b_m s^m}, \qquad a_i, b_i \in \mathcal{R},$$

it is clear that

$$
\begin{aligned}
\overline{H(j\omega)} \quad &= \quad \frac{\overline{a_0 + \cdots + a_n (j\omega)^n}}{\left(\overline{b_0 + \cdots + b_m (j\omega)^m}\right)} \\
\text{real coefficients} \quad &= \quad \frac{a_0 + \cdots + a_n (-j\omega)^n}{b_0 + \cdots + b_m (-j\omega)^m} \\
&= \quad H(-j\omega).
\end{aligned}
$$

Assuming that $|H_B(j\omega)|^2$ is truly the square of the magnitude of a transfer function with real coefficients, then we know that

$$
\begin{aligned}
|H_B(j\omega)|^2 &= H_B(j\omega)\overline{H_B(j\omega)} \\
&\overset{\text{real coefficients}}{=} H_B(j\omega)H_B(-j\omega) \\
&= H_B(s)H_B(-s)|_{s=j\omega}.
\end{aligned}
$$

Let us now look for $H_B(s)$ for $N = 1$. We find that

$$
|H_B(j\omega)|^2 = \frac{1}{1 + (j\omega/j\omega_c)^2}.
$$

That is,

$$
H_B(s)H_B(-s) = \frac{1}{1 + (s/j\omega_c)^2} = \frac{1}{1 - s^2/\omega_c^2} = \frac{1}{1 + s/\omega_c}\frac{1}{1 - s/\omega_c}.
$$

As we want a stable filter, we must make sure that all of the filters poles are in the left half-plane. Thus, we select

$$
H_B(s) = \frac{1}{1 + s/\omega_c}. \tag{23.1}
$$

This is the standard RC-type low-pass filter.

What happens when $N = 2$? We find that

$$
\begin{aligned}
H_B(s)H_B(-s) &= \frac{1}{1 + (s/j\omega_c)^4} \\
&= \frac{1}{1 + s^4/\omega_c^4} \\
&= \frac{1}{1 - js^2/\omega_c^2}\frac{1}{1 + js^2/\omega_c^2} \\
&= \frac{1}{1 - e^{j\pi/4}s/\omega_c}\frac{1}{1 + e^{j\pi/4}s/\omega_c}\frac{1}{1 - e^{3j\pi/4}s/\omega_c}\frac{1}{1 + e^{3j\pi/4}s/\omega_c} \\
&= \frac{1}{1 + \sqrt{2}s/\omega_c + s^2/\omega_c^2}\frac{1}{1 - \sqrt{2}s/\omega_c + s^2/\omega_c^2}.
\end{aligned}
$$

It is easy to see that if one selects

$$
H_B(s) = \frac{1}{1 + \sqrt{2}s/\omega_c + s^2/\omega_c^2} = \frac{\omega_c^2}{s^2 + \sqrt{2}s\omega_c + \omega_c^2},
$$

then one has a stable second-order Butterworth filter, and

$$
H_B(s)H_B(-s) = \frac{1}{1 + s^4/\omega_c^4}.
$$

23.3 Why We Always Succeed in Our Attempts at Factoring

Given a function that we thought might be the square of the absolute value of the frequency response of some filter, we have, so far, always been able to find a stable filter whose transfer function gives rise to just such a frequency response. Is this a coincidence? It is not.

We show that given a function $G(\omega)$

- that is a quotient of polynomials in ω^2 with real coefficients, $G(\omega) = P(\omega^2)/Q(\omega^2)$;
- that is bounded; and
- that is non-negative, $G(\omega) \geq 0$,

and letting $s = j\omega$, it is always possible to factor $G(s/j)$ as

$$G(s/j) = H(s)H(-s)$$

where all of the poles of $H(s)$ are located in the left half-plane (and all of the zeros of $H(s)$ are located in the left half-plane or on the imaginary axis). (Requiring that $G(\omega)$ be a function of ω^2 makes the magnitude response of the filter even—as is the case for filters whose impulse response is real.)

Assume that $P(\omega^2)$ and $Q(\omega^2)$ have no common factors. (If they initially have any common factors, cancel them and redefine $P(\omega^2)$ and $Q(\omega^2)$.) As $G(\omega)$ is bounded and the polynomials $P(\omega^2)$ and $Q(\omega^2)$ have no common factors, the function $Q(\omega^2)$ has no zeros for any real value of ω. $Q(\omega^2)$ never changes sign; we can pick its sign to be positive. As $G(\omega)$ is non-negative, we find that $P(\omega^2)$ is non-negative as well.

All polynomials with real coefficients can be factored into linear and quadratic factors such that the factors have real coefficients, and the quadratic factors are irreducible. Let us consider $Q(\omega^2)$ as a polynomial in ω^2. As $Q(\omega^2) > 0$, any linear factor in ω^2 must be of the form $\omega^2 + c^2, c > 0$. Let $s = j\omega$. Then, this factor can be written $-s^2 + c^2$. This in turn can be written

$$-s^2 + c^2 = (s + c)(-s + c) = r(s)r(-s),$$

and the lone zero of which $r(s)$ is possessed is located in the left half-plane. The remaining factors of $Q(\omega^2)$ are irreducible quadratics in ω^2—they are of the form

$$\omega^4 + a\omega^2 + b,$$

and, as the factors are irreducible quadratics in ω^2, $a^2 - 4b < 0$. As the polynomial is always positive, we find that $b > 0$. Thus, $|a| < 2\sqrt{b}$.

Letting $s = j\omega$, we find that the factor is of the form

$$s^4 - as^2 + b.$$

Let us see if this can be factored as we need—as

$$r(s)r(-s) = (s^2 + \alpha s + \beta)((-s)^2\alpha(-s) + \beta)$$

where the first factor, $r(s)$, has no zeros in the right half-plane. The product above is equal to

$$s^4 + (2\beta - \alpha^2)s^2 + \beta^2.$$

Equating coefficients with our original fourth-order polynomial, we find that $\beta^2 = b$ and $2\beta - \alpha^2 = a$. As we would like $r(s)$ to have all of its zeros in the right half-plane, the polynomial's coefficients must all have the same sign [6]. Thus, we let $\beta = +\sqrt{b}$. We find that $2\sqrt{b} - \alpha^2 = -a$. That is, $\alpha^2 = 2\sqrt{b} + a$. As we have shown that $2\sqrt{b} + a > 0$, the square root is real, and we find that $\alpha = \sqrt{2\sqrt{b} + a}$. We have shown that each factor of $Q(\omega^2)$ can be factored in precisely the fashion that we need.

$P(\omega^2)$ is a little harder to handle, because in addition to having terms of the types $Q(\omega^2)$ has, it *can* have zeros. Because $P(\omega^2)$ cannot change sign, any zeros of the form $\omega^2 - c$, $c > 0$ must have even multiplicity. Let us consider a single such double zero in ω^2, $(\omega^2 - c)^2$. Let us see if our new polynomial can be factored as needed. Letting $s = j\omega$, we find that we must consider $(s^2 + c)^2$. Noting that

$$(s^2 + c)^2 = r(s)r(-s) \equiv (s^2 + c)((-s)^2 + c),$$

we find that the new type of factor can itself be factored. The last possible type of zero is the polynomial ω^2 itself. As this function is non-negative, it need not appear with even multiplicity. Letting $\omega = s/j$, we find that

$$-s^2 = r(s)r(-s) = s(-s).$$

Here, too, we have succeeded in factoring the term of interest.

We have shown that given a function $G(\omega)$ that satisfies the list of properties above, one can *always* find a function $H(s)$ such that

$$H(s)H(-s) = G(s/j).$$

As the polynomials that make up the denominator of $H(s)$ have all of their roots in the left half-plane, the function $H(s)$ corresponds to a stable system. Additionally, the polynomials that make up the numerator can be chosen in such a way that all of their roots lie in the left half-plane or on the imaginary axis.

23.4 The Bilinear Transform

We have described several types of analog filters, and we have seen how to design Nth-order continuous-time Butterworth and Chebyshev filters. How can we take this knowledge and, without needing to think *too* much, convert it into knowledge about a *digital* filter?

We use the *bilinear transform*. The bilinear transform, which we develop momentarily, can be used to transform the transfer function of a stable analog filter into the transfer function of a similar stable digital filter.

The bilinear transform is defined by

$$z = B(s) \equiv \frac{1 + (T/2)s}{1 - (T/2)s}.$$

Its inverse is

$$s = B^{-1}(z) = \frac{2}{T}\frac{z-1}{z+1}.$$

This mapping is (essentially) a one-to-one and onto mapping of the complex plane into itself in such a way that $B(s)$ maps the imaginary axis onto the unit circle, the left half-plane into the interior of the unit circle and the right half-plane into the exterior of the unit circle [6]. One might say that it takes stable causal filters into stable causal filters. It is also easy to see that the bilinear transform takes Nth-order filters into Nth-order filters.

To see how the bilinear transform maps the unit circle, consider $s = B^{-1}(z)$ for $z = e^{Tj2\pi f}$. We find that

$$B^{-1}(z) = \frac{2}{T}\frac{e^{Tj2\pi f} - 1}{e^{Tj2\pi f} + 1}$$

$$= \frac{2}{T}j\tan(T2\pi f/2).$$

When Tf is small, this is approximately $j2\pi f$—which is the correct frequency in the "continuous-time domain."

Consider a continuous-time filter whose transfer function is $H(s)$ and whose frequency response is $H(j\omega)$. Making use of the bilinear transform to convert this filter into a discrete-time filter, we find that the transfer function of the discrete-time filter is given by $H(B^{-1}(z))$, and its frequency response is given by

$$H(B^{-1}(e^{j2\pi fT})) = H(2j\tan(T\pi f)/T).$$

Because when θ is small, $\tan(\theta) \approx \theta$, we find that as long as Tf is small,

$$H(B^{-1}(e^{j2\pi fT})) \approx H(2j\pi f) = H(j\omega).$$

That is, as long as f is small relative to the sampling frequency, the frequency response of the discrete-time filter is approximately the same as that of the continuous-time filter from which it was derived. (See Exercise 3 for information about prewarping—which is a way of dealing with the non-linearity in the mapping from the unit circle to the imaginary axis.)

23.5 The Passage from Analog Filter to Digital Filter

Let us consider the simple RC-type filter of (23.1)—the first-order Butterworth filter—and transform it into a digital filter. We replace every s with $B^{-1}(z)$. We find that

$$T(z) = \frac{1}{1 + B^{-1}(z)/\omega_c} = \frac{T\omega_c(z+1)}{T\omega_c(z+1) + 2(z-1)}.$$

The pole of this filter is located at

$$z = \frac{2 - T\omega_c}{2 + T\omega_c}.$$

If T is small enough, then this number is positive and less than one. When T is large, the pole is negative and less than one in absolute value. Additionally, $T(z = 1) = 1$—the system passes DC signals without change.

23.6 MATLAB and the Bilinear Transform

MATLAB® has a command, c2d, whose purpose is to convert continuous-time filters to discrete-time filters. The command has many options—as there are many ways of converting continuous-time filters to discrete-time filters.

Suppose that one would like to take a second-order, low-pass Butterworth filter whose cut-off frequency is 100 Hz, and convert it into a digital filter whose sampling rate is 1 kHz. On p. 158 we designed a second-order Butterworth filter. The program of Figure 23.2 defines the continuous-time transfer function, H, and then uses c2d to convert the filter to a discrete-time filter. Note that when used without a sample time, the command tf produces a continuous-time transfer function object (and MATLAB uses s to represent the transfer function's variable). The format of the c2d command, when used to convert a continuous-time transfer function object to a discrete-time transfer function object, is

c2d(continuous-time transfer function, sample time, 'tustin').

The third element of the command, the string 'tustin', tells MATLAB to make the conversion using the bilinear transform[2], rather than one of the other methods for converting continuous-time transfer functions to discrete-time transfer functions. The command hold on causes MATLAB to plot figures on the same set of axes. This makes it easy to use MATLAB to compare plots. Additionally, the command legend causes MATLAB to produce a legend for a figure.

The figure output by the sequence of commands given in Figure 23.2 is presented in Figure 23.3. Note that the frequency responses are substantially identical at low frequencies, but as the Nyquist frequency is approached the filters behave differently. This is precisely what we should have expected. At low frequencies the frequency response of the continuous-time filter is mapped into that of the discrete-time filter almost without change. As the frequency increases, the mapping starts to warp the response. That is why the responses differ at high frequencies.

---

[2] Arnold Tustin (1899–1994) introduced the bilinear transform that bears his name to the control community to relate discrete-time and continuous-time systems [2].

```
omega_c = 2 * pi * 100;
T = 0.0005;
H = tf([omega_c^2],[1 sqrt(2)*omega_c omega_c^2])
figure(1)
bodemag(H, '.-k', {2*pi, 2*pi*1000})
H_disc = c2d(H, T, 'tustin');
hold on
bodemag(H_disc , 'k', {2*pi, 2*pi*1000})
legend('continuous time', 'discrete time', 0)
```

Fig. 23.2. The MATLAB code for converting a continuous-time filter into a discrete-time filter using the bilinear transform

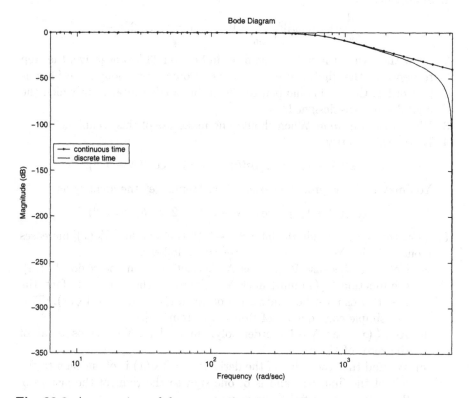

Fig. 23.3. A comparison of the magnitude response of a continuous-time filter and the discrete-time filter designed by transforming the continuous-time filter into a discrete-time filter by using the bilinear transform

23.7 The Experiment

Design a third-order analog Butterworth filter whose cut-off frequency is 135 Hz. Use the bilinear transform to convert the analog design into a digital one. Let $T_s = 1/1,350$ s. In the laboratory report, show all the computations. MATLAB may be used to help with the computations, but its built-in filter-design tools may not be used to help with the design. Implement the filter you design in C, and demonstrate that the filter works as predicted.

23.8 Exercises

1. Show that

$$\frac{d^n}{d\omega^n}|H_B(j\omega)|^2\bigg|_{\omega=0} = 0, \qquad n = 1, \ldots, 2N - 1.$$

2. Show that when the pole of the filter in Section 23.5 is negative, the step response of the digital filter is not monotonic even though the filter is first-order. Could this happen to the stable analog filter from which the digital filter was designed?

3. What is *prewarping*? When should one make use of this technique?

4. Derive the identity

$$\cos((N + 1)\theta) = 2\cos(\theta)\cos(N\theta) - \cos((N - 1)\theta).$$

You may find this problem easier if you "rephrase" the identity as

$$\cos((N + 1)\theta) + \cos((N - 1)\theta) = 2\cos(\theta)\cos(N\theta).$$

5. Show that for x outside the interval $[-1, 1]$, the function $|V_N(x)|$ increases monotonically. You may wish to proceed as follows:

   a) Note that because $V_N(x)$ has N distinct zeros in the region $[-1, 1]$, the function $V_N'(x)$ must have $N - 1$ zeros in that region. In fact, the $N - 1$ zeros must be sandwiched between the N zeros of $V_N(x)$. (This is a simple consequence of Rolle's theorem [17].)

   b) As $V_N'(x)$ is an $N - 1$th-order polynomial, these $N - 1$ zeros are all of the zeros of $V_N'(x)$.

   c) We find that the value of the derivative of $V_N(x)$ is of one sign to the left of the first zero and is of one sign to the right of the last zero. This allows us to finish the proof.

6. a) Calculate $H_C(s)$ for $N = 2$, $\epsilon = 1$, and $\omega_c = 1$.

   b) Make use of the bilinear transform to produce a second-order digital Chebyshev filter from the filter of the previous section. Let $T_s = 10$ ms.

   c) Examine the magnitude responses of the filters of the previous sections by making use of the MATLAB bodemag command. Compare the responses of the discrete-time and continuous-time systems.

7. Show that when $T\omega_c = 2$, the filter of Section 23.5 is an *FIR* low-pass filter.

24

New Filters from Old

Summary. In Chapter 23, we saw how to design low-pass digital filters using low-pass analog filters as a starting point. In this chapter, we describe one way of taking digital low-pass filters and converting them into other types of digital filters by transforming the filters' transfer functions according to simple rules.

Keywords. filter transformations, Blaschke products.

24.1 Transforming Filters

Let $H(z)$ be the transfer function of a stable filter, and let $G(z)$ be a function that

- is analytic for $|z| \leq 1$, and
- satisfies $|G(z)| = 1$ when $|z| = 1$.

What can be said about $H(G(z))$? As $G(z)$ maps points on the unit circle back to the unit circle, we can say that $G(z)$ "reorganizes" the frequency response of the filter. Functions like $G(z)$—which map the unit circle into itself—might possibly help us reorganize the frequency response of a low-pass filter into that of another type of filter.

24.2 Functions that Take the Unit Circle into Itself

It is easy to show that the absolute value of an analytic function, $G(z)$, in a domain is less than or equal to the absolute value of the function on the boundary of the domain—and this shows that if $G(z)$ takes the unit circle into itself, then inside the unit disk, $|G(z)| \leq 1$. The maximum modulus principle [3, p. 136] asserts that if the absolute value of an analytic function achieves its maximum inside a "reasonable domain," then the function is a constant within the domain. Given an analytic function $f(z)$ with no zeros inside a given domain, and considering the function $1/f(z)$, one can show that the minimum of the absolute value of the function in a "reasonable domain" must be greater than or equal to the minimum value of the absolute value on the boundary.

Let us consider a generic function, $G(z)$, that is analytic in the closed unit disk, $|z| \leq 1$, and that maps the unit circle into the unit circle. Because $G(z)$ maps the unit circle into the unit circle, we find that $|G(e^{j\theta})| = 1$. Thus, we find that inside the unit disk, $|G(z)| \leq 1$. If in addition $G(z)$ has no zeros in the unit disk, then we find that inside the unit disk $|G(z)| \geq 1$. This shows that if $G(z)$ has no zeros inside the unit disk, then $|G(z)| = 1$ throughout the unit disk. The maximum modulus principle then asserts that $G(z) = c$. In light of the fact that $|G(z)| = 1$, we find that $G(z) = e^{j\theta}$ (where $0 \leq \theta < 2\pi$).

This fact can be used to characterize all functions that are analytic inside the unit disk and that map the unit circle into itself. Let us consider a function of this sort, $G(z)$, that has a single zero of multiplicity one inside the unit disk at the point z_0. Consider the function

$$f(z) = G(z)\frac{1 - \bar{z}_0 z}{z_0 - z}.$$

Clearly, $f(z)$ has no zeros inside the unit disk—we have removed the only zero it had. The zero we have added is located at $z = 1/\bar{z}_0$, and $|z| > 1$. Let us consider how $f(z)$ maps the unit circle.

When $z = e^{j\theta}$, the second part of $f(z)$ satisfies

$$\frac{1 - \bar{z}_0 e^{j\theta}}{z_0 - e^{j\theta}} = \frac{1}{e^{j\theta}}\frac{1 - \bar{z}_0 e^{j\theta}}{z_0 e^{-j\theta} - 1}$$

$$= \frac{1}{e^{j\theta}}\frac{1 - c}{\bar{c} - 1}, \qquad c = z_0 e^{j\theta}.$$

The absolute value of this expression is always one. Thus, the function $f(z)$ is analytic, maps the unit circle into the unit circle, and *has no zeros inside the unit disk*. We conclude that $f(z) = e^{j\theta}$ (where θ is in no way related to the θ in the proof). We find that (at least inside and on the unit disk)

$$G(z) = e^{j\theta}\frac{z_0 - z}{1 - \bar{z}_0 z}.$$

As we know that $|G(z)| = 1$ on the unit circle, we know that $|G(z)| \leq 1$ in the closed unit disk—that $G(z)$ maps the unit disk into the unit disk. It is easy to show that $G(z)$ is a one-to-one and onto mapping of the complex plane into itself. (See Exercise 5.) As we know that $G(z)$ maps the unit circle into itself, maps the unit disk into itself, and that $G(z)$ is continuous, it is easy to show that $G(z)$ must also map the exterior of the unit disk into the exterior of the unit disk.

The same trick that we used to remove a single zero can be used to remove as many zeros as the function has. Thus, all functions that are analytic inside and on the unit disk, that have a finite number of zeros inside the unit disk, and that map the unit circle into itself must be of the form

$$e^{j\theta} \prod_{k=1}^{N} \frac{z_k - z}{1 - \bar{z}_k z}, \qquad |z_k| < 1. \tag{24.1}$$

(Such functions are known as finite Blaschke products [18].) It is easy to see that such functions map the exterior of the unit disk to the exterior of the unit disk.

Let $G(z)$ be such a function, let $H(z)$ be the transfer function of a stable filter, and consider $H(G(z))$. Since $G(z)$ maps the exterior of the unit disk into itself, if $|z| > 1$, then $|G(z)| > 1$. As $H(z)$ corresponds to a stable filter, it has no poles outside the unit disk. Thus, $H(G(z))$ cannot be infinite if z is outside the unit disk. We find that $H(G(z))$ has no poles outside the unit disk and is, consequently, the transfer function of a stable filter. Because $G(z)$ is a rational function of z, $H(G(z))$ is as well. As $G(z)$ takes points on the unit circle to points on the unit circle, the frequency response of the filter whose transfer function is given by $H(G(z))$ is a "rearranged" version of the frequency response of the original filter.

In the following sections, we make use of finite Blaschke products to rearrange the frequency response of a digital filter so that it becomes a different type of digital filter. (A more complete set of filter transformation rules can be found in [4].)

24.3 Converting a Low-pass Filter into a High-pass Filter

The first of the mappings—of the reorganizations—that we consider is the mapping for which $N = 1$, $z_1 = 0$, and $e^{j\theta} = +1$. That is, we consider the mapping $-z$. It is easy to see that under this mapping, low frequencies—which correspond to values of z near 1—go over to high frequencies—which correspond to values of z near -1.

Consider the simple low-pass filter

$$G_{lp}(z) = (1 - \alpha)\frac{z}{z - \alpha}, \qquad 0 < \alpha < 1.$$

Under our mapping, this becomes

$$G_{hp}(z) = (1 - \alpha)\frac{z}{z + \alpha}, \qquad 0 < \alpha < 1.$$

It is easy to show that this is indeed a high-pass filter. Using MATLAB® and its **bodemag** command, it is even easier to see that this is a high-pass filter.

Consider, for example, the filter for which $\alpha = 1/2$ and $T_s = 1$ ms. Giving MATLAB the commands

```
alpha = 0.5;
G = tf([(1 - alpha) 0],[1 alpha], 0.001)
bodemag(G)
```

causes MATLAB to produce the plot shown in Figure 24.1. This plot is clearly the magnitude response of a high-pass filter.

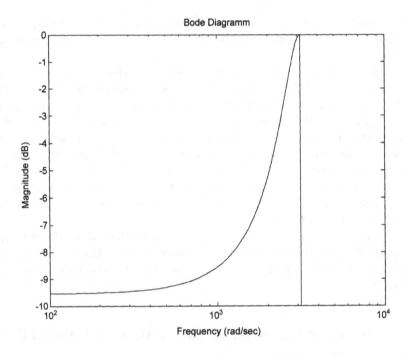

Fig. 24.1. The magnitude response of a high-pass filter

24.4 Changing the Cut-off Frequency of an Existing Low-pass Filter

We would like a function of the form (24.1) that maps the unit circle into itself in such a way that the frequencies near zero are changed. We would like to "stretch" that range.

It is not hard to see that N in (24.1) controls how many times the unit circle is mapped onto itself. In our case, we want it to be mapped onto itself once—we just want to stretch it a bit. Thus, we take $N = 1$. As we would like to preserve the realness of the coefficients of the final transfer function, we require that $e^{j\theta} = \pm 1$ and $z_1 \in \mathcal{R}$. As $|z_1| < 1$, this leaves us with the set of mappings

$$T_{z_1}(z) \equiv \pm \frac{z_1 - z}{1 - z_1 z}, \qquad -1 < z_1 < 1.$$

Note that $T(1) = \pm(-1)$. If we are looking to take low-pass filters into low-pass filters, we must limit ourselves to filters of the form

$$T_{z_1}(z) \equiv -\frac{z_1 - z}{1 - z_1 z} = \frac{z - z_1}{1 - z_1 z}.$$

Such filters will change the cut-off frequency of a low-pass filter. To obtain a high-pass filter with a specific cut-off frequency, one would compose the two mappings $-z$ and $T_{z_1}(z)$; one would make use of the mapping $-T_{z_1}(z)$.

Let us consider the action of the mappings

$$T_a(z) = \frac{z - a}{1 - az}, \qquad -1 < a < 1$$

in greater detail. We find that this mapping takes the point $z = e^{2\pi j f T_s}$ to the point

$$T_a(e^{2\pi j f T_s}) = e^{2\pi j f T_s} \frac{1 - ae^{-2\pi j f T_s}}{1 - ae^{2\pi j f T_s}} = e^{2\pi j f T_s} \frac{1 - ae^{-2\pi j f T_s}}{\left(1 - ae^{-2\pi j f T_s}\right)}.$$

We know that our mappings take the unit circle into the unit circle. Thus, the magnitude of this point must be one. The phase of the point is

$$\angle T_a(e^{2\pi j f T_s}) = 2\pi f T_s + 2\tan^{-1}\left(\frac{a\sin(2\pi f T_s)}{1 - a\cos(2\pi f T_s)}\right).$$

We find that the mapping shifts lower frequencies to higher ones if a is positive, and it shifts higher frequencies to lower ones if a is negative.

Let us consider an example. Suppose that we have a low-pass filter that passes frequencies up to 1 Hz, and we would like a low-pass filter that passes frequencies up to 10 Hz. We can make use of our transformation to design a function that will transform our filter appropriately.

We are looking for an a for which an input of 10 Hz is converted into the angle that is appropriate to 1 Hz. Thus, we must solve the equation

$$2\pi 1 T_s = 2\pi 10 T_s + 2\tan^{-1}\left(\frac{a\sin(2\pi 10 T_s)}{1 - a\cos(2\pi 10 T_s)}\right).$$

Let us take $T_s = 1\,\text{ms}$—let us consider a filter that samples 1,000 times per second. We find that we are now looking for the solution of the equation

$$\tan(-0.001 \times 9\pi) = \frac{a\sin(2\pi 0.01)}{1 - a\cos(2\pi 0.01)}.$$

Solving for a, we find that

$$a = -0.8182.$$

Thus, the mapping

$$T_a(z) = \frac{z + 0.8182}{1 + 0.8182z}$$

can be used to convert a low-pass filter that takes 1,000 samples per second, and whose cut-off frequency is 1 Hz, into a low-pass filter whose cut-off frequency is 10 Hz.

24.5 Going from a Low-pass Filter to a Bandpass Filter

We have seen that the transforms $T_{z_1}(z), -1 < z_1 < 1$, take low-pass filters into other low-pass filters by traversing the unit circle at different "rates." If one takes two such transforms and multiplies them, one causes the unit circle to be traversed once as f goes from zero to $1/(2T_s)$. Such a transformation causes the low-pass filter to become a bandstop filter. If one uses the transformation on a high-pass filter, the high-pass filter becomes a bandpass filter.

Suppose that $T_{LP}(z)$ is the transfer function of a low-pass filter. We have found that $T_{LP}(-T_\alpha(z)T_\beta(z))$ is a bandpass filter. Suppose that $T_{LP}(z)$ is a low-pass filter with cut-off frequency F_c, and that we let $\alpha = \beta = 0$—we choose $T_\alpha(z) = T_\beta(z) = z$. Let us characterize the filter given by

$$T(z) = T_{LP}(-z^2).$$

Substituting $z = e^{2\pi j f T_s}$ in $T(z)$, we find that

$$T(e^{2\pi j f T_s}) = T_{LP}(-e^{4\pi j f T_s}) = T_{LP}(e^{j2\pi(2f - F_s/2)T_s}).$$

When $f = 0$, the frequency response of $T(z)$ corresponds to the response of the low-pass filter when $f = -F_s/2$. This will be very small—as seen in Figure 24.2. (Figure 24.2 is the magnitude response of a typical low-pass filter. In the sample system, $F_s = 100$ Hz, and we have plotted the magnitude response from $-1/(2T_s)$ to $1/(2T_s)$.) As f increases, the magnitude of the frequency response increases. When $f = F_s/4 - F_c/2$, the frequency response of the new filter corresponds to that of the low-pass filter at the (left-hand) cut-off frequency. When f reaches $f = F_s/4$, the response corresponds to that of the low-pass filter at $f = 0$. As f increases further, the response of the filter starts to "fall off." At $f = T_s/4 + F_c/2$, the filter has once again reached the low-pass filter's cut-off frequency. When $f = 1/(2T_s)$, the response of the new filter corresponds to that of the low-pass filter at $f = F_s/2$—and should, once again, be very small. We find that the new filter is a (symmetric) bandpass filter, and its cut-off frequencies are $F_s/4 - F_c/2$ and $F_s/4 + F_c/2$.

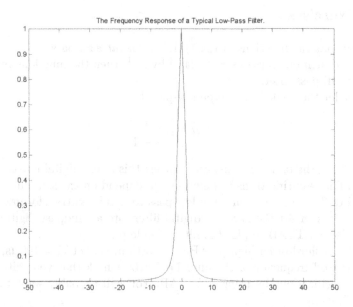

Fig. 24.2. A typical low-pass filter. The far left of the plot is $-F_s/2$, and the far right is $F_s/2$.

24.6 The Experiment

1. Let $G(z) = (1 + z^{-1} + z^{-2})/3$.
2. Use this low-pass prototype to design a high-pass, a bandstop, and a bandpass filter.
3. Implement the filters using Simulink®, and test their frequency responses. Let $T_s = 1\,\text{ms}$.

24.7 The Report

Write a report that explains each step of the design process. Include the Bode plots of the filters that were implemented. Finally, include some plots of the input to, and output of, the system that show that the filters that were designed really have the frequency response indicated by the Bode plots.

24.8 Exercises

1. Show that the functions of (24.1) define *unstable* all-pass.
2. Show that if one replaces z of (24.1) by z^{-1}, then the functions represent *stable* all-pass filters.
3. Consider the analog prototype low-pass filter

$$G(s) = \frac{1}{s+1}.$$

   a) Using the bilinear transform, convert this into a digital low-pass filter. (The resulting transfer function will depend on the sampling rate.)
   b) Let $T_s = 100\,\mu s$. Using the low-pass to bandpass transformation $z \rightarrow -z^2$, convert the low-pass digital filter into a bandpass digital filter.
   c) Use MATLAB to plot the filter's Bode plots.
4. Design a third-order high-pass Butterworth filter. Let $T_s = 100\,\mu s$, and let the cut-off frequency of the filter be 4 kHz. Check that your filter meets the specifications by using MATLAB. You will probably want to design the filter by
   a) Designing a continuous-time low-pass Butterworth filter with a cut-off frequency of 1 Hz.
   b) Modifying the filter to have a cut-off frequency that will lead to a final design with a cut-off frequency of 4 kHz. (How does one find the cut-off frequency of the low-pass filter?)
   c) Converting the continuous-time filter into a discrete-time filter by making use of the bilinear transform.
   d) Converting the low-pass filter into a high-pass filter.
5. Show that the function
$$K(z) = \frac{z_k - z}{1 - z\bar{z}_k}$$

satisfies the equation $K(K(z)) = z$. Explain how this shows that the function $K(z)$ is a one-to-one and onto mapping of the complex plane into itself. As $G(z) = e^{j\theta}K(z)$, conclude that $G(z)$ is also a one-to-one and onto mapping of the complex plane into itself.

25

Implementing an IIR Digital Filter

Summary. We now understand how to "design" a recurrence relation that implements a digital filter. In this chapter, we consider how such equations should be implemented. We discuss several methods of implementing digital filters, and we find that due to numerical stability issues, it is best to implement high-order filters by using *biquads*.

Keywords. direct form I realization, direct form II realization, numerical stability, biquads.

25.1 Introduction

As the digital filters that are of interest to us can be described by transfer functions that are rational functions of z^{-1}, it is easy to show that the filters can be described by equations of the form

$$y_n = \sum_{m=1}^{M} b_m y_{n-m} + \sum_{l=0}^{L} a_l x_{n-l}. \tag{25.1}$$

Translating this into a transfer function, we find that the transfer function of a system described by (25.1) is

$$T(z) = \frac{Y(z)}{X(z)} = \frac{\sum_{l=0}^{L} a_l z^{-1}}{1 - \sum_{m=1}^{M} b_m z^{-m}}.$$

There are many possible ways that one can actually calculate the sums of (25.1). The simplest is to perform the calculations as given—in one fell swoop. We consider this method in Section 25.2. In later sections, we examine more interesting and effective techniques.

25.2 The Direct Form I Realization

Making use of block diagram notation, and noting that a block whose transfer functions is z^{-1} delays a sample by one sample period, we find that one way of expressing the calculation that we would like to perform is to use the block diagram of Figure 25.1. This realization of the calculation, which is the most intuitive of all the realizations, is known as the direct form I realization of the system. Note that in this realization we need to store L delayed samples of the input and M delayed samples of the output. As we see in Section 25.3, by simply rewriting the equations it is possible to do quite a bit better.

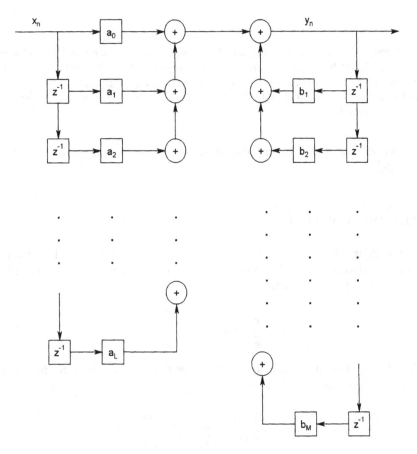

Fig. 25.1. The block diagram corresponding to the direct form I realization

25.3 The Direct Form II Realization

By reconsidering the transfer function, it is possible to substantially reduce the number of samples that need to be stored. Let us define the auxiliary variable w_n by the equation

$$W(z) = \frac{1}{1 - \sum_{m=1}^{M} b_m z^{-m}} X(z),$$

and note that with this definition we find that $Y(z)$ satisfies

$$Y(z) = \sum_{l=0}^{L} a_l z^{-l} W(z).$$

Both w_n and y_n depend on previous values of w_n but *not* on previous values of either y_n or x_n. The block diagram of a w_n-based system is given in Figure 25.2. We find that the number of delays necessary here—and the number of elements that must be stored here—is the larger of L and M. If these numbers are roughly equal, then this method reduces the number of terms that must be stored by nearly half. This realization of the calculation is known as the direct form II realization. In Figure 25.2, the last of the a_l is given as a_M, rather than as a_L. This is done because generally $L \leq M$—because the transfer function is generally proper. If $L < M$, we let

$$a_l = 0, \qquad l = L+1, \ldots M.$$

In this way, the figure correctly and simply represents the calculations that we need to perform.

25.4 Trouble in Paradise

It turns out that both of the realizations we have seen so far are problematic when M or L is even reasonably large. What is the problem? Generally speaking, the coefficients of the polynomials in the transfer function are implemented in some type of finite precision arithmetic. That means that the coefficients used in the actual calculations are not precisely the coefficients that were planned. Perturbing the coefficients of a polynomial perturbs the location of the polynomial's roots. The roots of the denominator are the poles of the system, and if the poles leave the unit disk, then the system becomes unstable.

Let us see how perturbing the coefficients of a polynomial affects the placement of the roots of the polynomial. Consider a polynomial of the form

$$P(z) = 1 - \sum_{m=1}^{M} b_m z^{-m}.$$

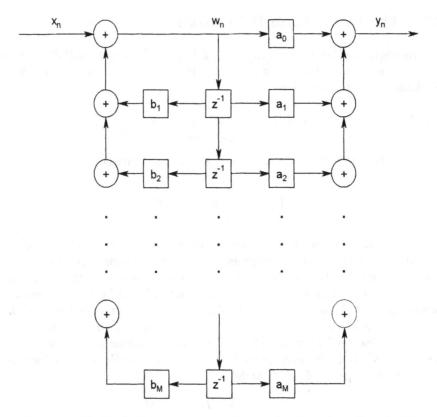

Fig. 25.2. The block diagram corresponding to the direct form II realization

The roots of the polynomial are the solutions of the equation $P(z) = 0$. Let us consider a particular root, z_k, as a function of the coefficient b_i, and let us emphasize this by writing the equation satisfied by the root as $P(z_k, b_i) = 0$. Differentiating with respect to b_i and making use of the chain rule, we find that

$$\frac{\mathrm{d}}{\mathrm{d}b_i} P(z_k, b_i) = \frac{\partial}{\partial z_k} P(z_k, b_i) \frac{\partial z_k}{\partial b_i} + \frac{\partial}{\partial b_i} P(z_k, b_i) \frac{\partial b_i}{\partial b_i}$$

$$= P'(z_k) \frac{\partial z_k}{\partial b_i} - z_k^{-i} = 0$$

where

$$P'(z_k) = \left. \frac{\mathrm{d}}{\mathrm{d}z} P(z) \right|_{z=z_k}.$$

Thus, we find that

$$\frac{\partial z_k}{\partial b_i} = \frac{z_k^{-i}}{P'(z_k)}.$$

Note that $P(z)$ can also be written in the form

$$P(z) = \prod_{m=1}^{M} (1 - z^{-1}z_m).$$

In this form, it is clear that

$$P'(z) = \sum_{n=1}^{M} \prod_{m=1,m \neq n}^{M} (1 - z^{-1}z_m)(z^{-2}z_n).$$

Evaluating this sum at the root z_k, we find that

$$P'(z_k) = \prod_{m=1,m \neq k}^{M} (1 - z_k^{-1}z_m)z_k^{-1}.$$

Combining all of our results, we find that, at a root,

$$\frac{\partial z_k}{\partial b_i} = \frac{z_k^{-i}}{\prod_{m=1,m \neq k}^{M}(1 - z_k^{-1}z_m)z_k^{-1}} = \frac{z_k^{M-i}}{\prod_{m=1,m \neq k}^{M}(z_k - z_m)}.$$

We find that if the roots of the polynomial are near one another, this value can be quite large, and even a small change in the coefficients of the polynomial can lead to a large change in the location of the polynomial's roots.

25.5 The Solution: Biquads

We have seen that if one tries to implement a filter by using a single section—as is done in both the realizations that we have seen—the implementation may not perform as expected. In particular, if the poles or zeros of the filter are clustered, then small inaccuracies in the coefficient values may be translated into large inaccuracies in the pole or zero locations. As the poles of a stable filter are all located inside the unit disk, a filter with many poles will have some poles that are located near one another. This will cause the poles' positions to be very sensitive to small changes in the filter coefficients. This sensitivity problem can cause a filter that should—theoretically—have been stable to be unstable in practice.

One way to overcome this problem is to decompose the filter's transfer function into a product of transfer functions whose poles and zeros are well separated. In that way inaccuracies in the values of the coefficients of the individual filters will not affect the poles and zeros of the system too much. The standard way to rewrite the system's transfer function is as a product of second-order sections—of sections whose numerator and denominator polynomials are second order. Such sections are generally referred to as *biquads*. The block diagram of a filter that has been decomposed into cascaded biquads is given in Figure 25.3.

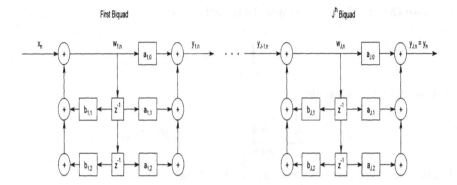

Fig. 25.3. The block diagram corresponding to the biquad realization. Here, the direct form II realization is used.

25.6 Exercises

1. Determine the recurrence relation satisfied by x_n and y_n of Figure 25.4.

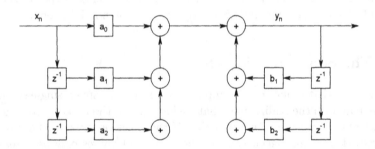

Fig. 25.4. A block diagram of a single biquad. Here, the direct form I realization is used.

2. Recall that a polynomial with real coefficients can always be factored into first- and second-order polynomials with real coefficients. Use this fact to explain why high-order filters are generally decomposed into second-order sections, and not first-order sections.

3. Consider the polynomial $P(z) = z^2 - 2\cos(\theta)z + 1$.
   a) Show that when $|\theta| << 1$, the polynomial has two nearly identical roots.
   b) Show that when $|\theta - \pi/2| << 1$, the roots are well separated.
   c) Show that when θ is near zero, small changes in $\cos(\theta)$ lead to relatively large changes in the location of the roots of the polynomial.

d) Finally, show that when θ is near $\pi/2$, small changes in $\cos(\theta)$ do not affect the location of the roots as much as they did in the previous case.

4. a) Design a fourth-order continuous-time Butterworth low-pass filter with a cut-off frequency of $w_c = 10\,\text{rad}\,\text{s}^{-1}$.
   b) Using the bilinear transform, transform this filter into a digital filter for which $T_s = 0.001\,\text{s}$.
   c) Find the recurrence relation satisfied by the filter's input and output.
   d) Give a schematic drawing (with all the filter coefficients) of the filter implemented as a cascade of two biquads realized using the direct form II realization.

26

IIR Filter Design Using MATLAB

Summary. Though it is possible to use the beautiful and interesting methods of days gone by to design analog or IIR digital filters, it is not strictly necessary. MATLAB® has two different "levels" of filter-design tools. It has "regular" commands for filter design, and it has a graphical user interface (GUI) based filter-design tool as well. We discuss both options in the following sections.

Keywords. MATLAB, butter, fdatool.

26.1 Individual Commands

MATLAB's suite of filter-design commands includes commands to design Butterworth, Chebyshev, and elliptic filters. As is generally the case when using MATLAB, one can find out how things work by using the MATLAB help command.

We consider the simplest command—butter. To design an Nth-order digital filter with cut-off frequency ω_n, one writes [b a] = butter(N, Wn). The arrays b and a are the coefficients of the polynomials in the filter's numerator and denominator, respectively.

When designing a digital filter, all frequencies can be considered as being relative to the sampling frequency. That is, if one is sampling at 100 samples s^{-1} and one's cut-off frequency is 10 Hz, one can intelligently say that one's cut-off frequency is one-tenth the sampling frequency. In fact, if one "speeds up" the sampling rate to 1,000 samples s^{-1} and one uses the same filter—a filter that uses the same formula in its implementation—the cut-off frequency will now be one-tenth of the new sampling rate. That is, the cut-off frequency will be 100 Hz. (See Exercise 2.)

When using the butter command, the frequency is given relative to the Nyquist frequency—relative to half the sampling frequency. The number Wn must be between 0 and 1. The cut-off frequency will be Wn times the Nyquist frequency of the filter that is being implemented.

Let us consider a simple example. Let us design and examine a Butterworth low-pass filter whose sampling frequency is $200\,\text{samples}\,\text{s}^{-1}$ and whose cut-off frequency is $20\,\text{Hz}$. As the Nyquist frequency is $100\,\text{Hz}$, we find that Wn = $20/100 = 0.2$. Thus, our first step is to use the **butter** command as follows. (Both the command and MATLAB's response are given.)

```
>> [b a] = butter(3, 0.2)

b =

    0.0181    0.0543    0.0543    0.0181


a =

    1.0000   -1.7600    1.1829   -0.2781
```

In order to examine this design, we implement it as a discrete-time transfer function. To do this, we use the **tf** command. This command takes a numerator, a denominator, and a sample time and returns a "transfer function object." Our command and MATLAB's response follow.

```
>> tf(b, a, 0.005)

Transfer function:
0.0181 z^3 + 0.0543 z^2 + 0.0543 z + 0.0181
-------------------------------------------
    z^3 - 1.76 z^2 + 1.183 z - 0.2781

Sampling time: 0.005
```

Now one can use all of MATLAB's standard commands to explore how this filter behaves. One might use **bode** to see the filter's Bode plots, or **step** to examine its step response. From the transfer function it is also easy to construct the recurrence relation that defines the digital filter.

In our case, for example, after dividing the numerator and the denominator by z^3, we find that

$$\frac{Y(z)}{X(z)} = \frac{0.0181 + 0.0543z^{-1} + 0.0543z^{-2} + 0.0181z^3}{1 - 1.76z^{-1} + 1.183z^{-2} + 0.2781z^{-3}}$$
$$\Leftrightarrow y_k = 1.76y_{k-1} - 1.183y_{k-2} + 0.2781y_{k-3}$$
$$+0.0181x_k + 0.0543x_{k-1} + 0.0543x_{k-2} + 0.0181x_{k-3}.$$

26.2 The Experiment: Part I

Use the built-in MATLAB commands to design a second-order Butterworth low-pass filter. Design the filter to have a cut-off frequency of 270 Hz when the sampling rate is 2,700 samples s^{-1}. Implement the filter using the ADuC841. Measure the frequency response of the filter at several frequencies and compare the results with those predicted by the theory.

When implementing the filter, consider carefully what data type(s) to use in your C program. Make sure that, on the one hand, you have enough bits for your needs. On the other hand, make sure that you do not make the microprocessor's job too hard. Floating point arithmetic is rather hard on simple microprocessors like the ADuC841.

26.3 Fully Automatic Filter Design

In addition to the "discrete filter-design commands," MATLAB also provides a "filter design and analysis tool." This tool is invoked by typing `fdatool` at the MATLAB command line. This tool will design most any filter. It asks for your specifications and returns the Bode plots of the filter it designs. This tool has many display and implementation options. Play with it a little bit to see what it can do!

26.4 The Experiment: Part II

Use `fdatool` to design a third-order low-pass elliptic filter. Let the cut-off frequency of the filter be 270 Hz and the sampling frequency be 2,700 samples per second. Let the passband ripple—the extent to which the magnitude is allowed to vary in the passband—be 1 dB, and make the magnitude in the stopband at least 30 dB below the passband magnitude.

Note that MATLAB uses second-order sections—biquads—in its default filter implementation. For a third-order filter, one probably does not need to use this type of implementation. To change the implementation type, right-click on the "Current Filter Information" box, and click on "convert to single section." In order to see the filter coefficients, either click on the button labeled "[b/a]" in the toolbar or go to the File tab, click on "Export...," and tell it to export as coefficients. After this procedure, you will find variables in your MATLAB workspace that contain the filter coefficients.

Implement the filter using the ADuC841. Measure the frequency response of the filter at several frequencies, and compare the results with those predicted by the theory.

26.5 Exercises

1. a) Use the **fdatool** to design a sixth-order elliptic low-pass filter. Let the cut-off frequency be 1 kHz and the sampling frequency be 10 kHz. For the elliptic filter, let the passband attenuation, **Apass**, be 1 dB and the stopband attenuation, **Astop**, be 30 dB.
   b) Compare the filter of the preceding section with a Butterworth filter with the same cut-off frequency and sampling rate.
   c) Which filter has a narrower transition region?
   d) Which filter has a smoother frequency response?
2. Explain why doubling the sampling rate of a digital filter that is described by a recurrence relation doubles the values of all of a filter's frequency-related parameters. You may want to consider the filter's frequency response for a sampling rate of T_s and for a sampling rate of $T_s/2$.

Group Delay and Phase Delay in Filters

Summary. In this chapter, we explore the way in which a filter's output is delayed with respect to its input. We show that if the filter's phase is not linear, then the filter delays signals with different carrier frequencies by different amounts.

Keywords. group delay, phase delay, bandpass signals, carrier frequency.

27.1 Group and Phase Delay in Continuous-time Filters

Suppose that one has a narrow-band bandpass signal, $y(t)$. That is, suppose that $y(t)$ can be written as

$$y(t) = e^{2\pi j F_c t} x(t)$$

where $X(f) = 0$ for all $|f| > B$, $B << F_c$, and F_c is the signal's carrier frequency.

Consider the Fourier transform representation of $x(t)$

$$x(t) = \int_{-B}^{B} e^{2\pi j f t} X(f)\, df$$

$$= \int_{-B}^{B} |X(f)| e^{j \angle X(f)} e^{2\pi j f t}\, df.$$

We find that $x(t)$ is composed of sinusoids of the form

$$|X(f)| e^{j \angle X(f)} e^{2\pi j f t}, \quad |f| \le B.$$

Similarly, we find that $y(t)$ is composed of sinusoids of the form

$$|X(f)| e^{j \angle X(f)} e^{2\pi j (f + F_c) t}, \quad |f| \le B.$$

When the components of the function $y(t)$ pass through a filter whose frequency response is $H(f)$, then the components of the resulting function, which we call $z(t)$, are

$$|H(F_c + f)|e^{j\angle H(F_c+f)}|X(f)|e^{j\angle X(f)}e^{2\pi j(f+F_c)t}, \quad |f| \le B.$$

Let us define the function $\Theta_H(f) \equiv \angle H(f)$. For relatively small f we can approximate $\Theta_H(F_c + f)$ by $\Theta_H(F_c) + \Theta'_H(F_c)f$. Also, assuming that the magnitude of the filter response does not change much for small f, we can approximate $|H(F_c + f)|$ by $|H(F_c)|$. (This "0th-order" approximation is reasonable, because a small enough error in the magnitude can only produce a small error in the final estimate of the signal.) We approximate the constituent components of $z(t)$ by

$$|H(F_c)|e^{j(\Theta_H(F_c)+\Theta'_H(F_c)f)}|X(f)|e^{j\angle X(f)}e^{2\pi j(f+F_c)t}, \quad |f| \le B.$$

Rewriting this, we find that the constituent components of $z(t)$ are

$$|H(F_c)|e^{j\Theta_H(F_c)}e^{2\pi jF_ct}|X(f)|e^{j\angle X(f)}e^{j\Theta'_H(F_c)f}e^{2\pi jft}, \quad |f| \le B.$$

This, in turn, can be written as

$$|H(F_c)|e^{2\pi jF_c(t+\Theta_H(F_c)/(2\pi F_c))}|X(f)|e^{j\angle X(f)}e^{2\pi jf(t+\Theta'_H(F_c)/(2\pi))}, \quad |f| \le B.$$

This, however, shows that

$$z(t) \approx \int_{-B}^{B} |H(F_c)|e^{2\pi jF_c(t+\Theta_H(F_c)/(2\pi F_c))}|X(f)|e^{j\angle X(f)}e^{2\pi jf(t+\Theta'_H(F_c)/(2\pi))} \, df$$

$$= |H(F_c)|e^{2\pi jF_c(t+\Theta_H(F_c)/(2\pi F_c))} \int_{-B}^{B} X(f)e^{2\pi jf(t+\Theta'_H(F_c)/(2\pi))} \, df$$

$$= |H(F_c)|e^{2\pi jF_c(t+\Theta_H(F_c)/(2\pi F_c))}x(t + \Theta'_H(F_c)/(2\pi)).$$

We find that the *carrier* has been delayed by

$$\text{phase delay} \equiv -\Theta_H(F_c)/(2\pi F_c), \tag{27.1}$$

and the envelope, the signal $x(t)$, has been delayed by

$$\text{group delay} \equiv -\Theta'_H(F_c)/(2\pi). \tag{27.2}$$

The group delay is constant precisely when $\Theta_H(f)$ is a linear function of the frequency. One example of a filter with constant group delay is a symmetric FIR filter. See Chapter 28 for more details about this important class of filters.

27.2 A Simple Example

Let us consider the first-order filter whose transfer function is

$$H(s) = \frac{1}{s+1}$$

and whose frequency response is

$$H(2\pi jf) = \frac{1}{2\pi jf + 1}.$$

Clearly,

$$\Theta_H(f) = -\tan^{-1}(2\pi f).$$

We find that the group delay is

$$\text{group delay} = -\frac{\Theta'(f)}{2\pi} = \frac{1}{(2\pi f)^2 + 1}.$$

We find that the group delay is largest when $f = 0$; at that point, the group delay is $1\,\text{s}$.

If we would like to "see" this delay, we must pass a signal whose carrier frequency is $0\,\text{Hz}$—an unmodulated signal—through the filter. We must be careful to see to it that the signal's energy is contained in frequencies for which the phase response of the filter is reasonably linear. The Bode plots corresponding to $H(2\pi jf)$ are given in Figure 27.1. We note that the phase response is rather linear at frequencies less than, or equal to, $0.2\,\text{Hz}$. Thus, we need a very low-frequency signal. We make use of the signal

$$x(t) = \frac{\sin(0.2\pi(t - 10))}{0.2\pi(t - 10)}.$$

The output of the filter (as generated by Simulink® and displayed by MATLAB®) is given in Figure 27.2. Note that the output is delayed from the input by $1\,\text{s}$—as predicted.

27.3 A MATLAB Experiment

1. Write a short MATLAB function to calculate the group delay and gain associated with a transfer function. The function should accept a transfer function object (created using the tf command) and a frequency. The function should return the gain and group delay at the frequency. (You may find that when used properly, the MATLAB command bode is quite helpful here.)

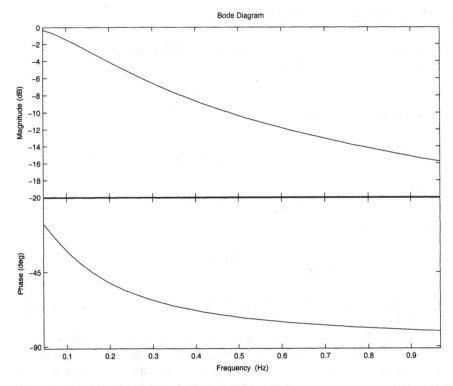

Fig. 27.1. The Bode plots of the filter under consideration. Note that the phase response is rather linear for frequencies below 0.2 Hz.

2. Create a Simulink model in which the group delay caused by a system can be observed. You may want to use a system whose transfer function is

$$G(s) = \frac{\omega_0 \epsilon}{s^2 + \epsilon s + \omega_0^2}.$$

Use a modulated waveform as the system's input, and explain how one sees the effects of the group delay in the system's output.

3. Using the results of the program written in the first section, explain the results of the second section.

27.4 Group Delay in Discrete-time Systems

There is no real difference between the results for discrete-time and continuous-time systems. The only difference is that $H(f)$ must be defined as the frequency response of the discrete-time system. Thus, if the transfer function of the system is $H(z)$, then

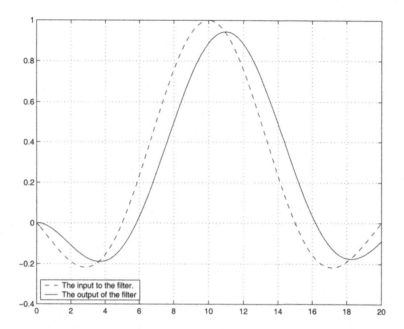

Fig. 27.2. The input to the filter, and the output of the filter. The output is delayed from the input by about 1 s.

$$\Theta_H(f) \equiv \angle H(e^{2\pi j f T_s})$$

where T_s is the sampling period.

27.5 Exercises

1. a) Calculate the group delay function that corresponds to the filter whose transfer function is
$$H(s) = \frac{s/100 + 1}{s/1000 + 1}.$$
   b) For which frequencies is the group delay negative?
2. Calculate the group delay function that corresponds to the filter whose transfer function is
$$H(s) = \frac{\omega_0 \epsilon}{s^2 + 2\epsilon s + \omega_0^2}.$$
Remember to "mind your Ps and Qs" when calculating angles.
3. Calculate the group delay function that corresponds to the filter whose transfer function is
$$H(z) = \frac{T_s z}{z - 1}.$$
You need only consider the frequencies $0 < f < 1/(2T_s)$. Remember to "mind your Ps and Qs" when calculating angles.

Design of FIR Filters

Summary. In Chapter 22, we considered two simple finite impulse response (FIR) filters. We have not yet considered how one goes about *designing* a generic finite impulse response filter. We now remedy that lack.

Keywords. FIR filter design, symmetric FIR filter, windows, Fourier series, group delay.

28.1 FIR Filter Design

One way of designing a filter is to take the desired frequency response and find the precise set of coefficients that gives one such a filter. As we will see, the coefficients that correspond to the most commonly desired filters cannot be implemented "as is."

Suppose that one chooses to sample one's input T_s times per second and that one would like a filter $H(z)$ whose frequency response is $\tilde{H}(f)$ (between $f = -F_s/2$ and $f = F_s/2$). That is, one would like

$$H(e^{2\pi j f T_s}) = \tilde{H}(f).$$

Generally speaking, we start by requiring that $\tilde{H}(f)$ be real and even. (These properties guarantee that the impulse response of the filer is real and even. See Exercise 3.) As $\tilde{H}(f)$ is supposed to be the frequency response of a discrete-time filter, it must be periodic with period $F_s = 1/T_s$ (as we saw in Section 18.10).

Assuming that

$$H(z) = \sum_{m=-\infty}^{\infty} h_m z^{-m},$$

we find that we are looking for coefficients, h_m, that satisfy

$$\sum_{m=-\infty}^{\infty} h_m e^{-2\pi jmfT_s} = \tilde{H}(f).$$

The h_m are essentially the coefficients of the Fourier series associated with $\tilde{H}(f)$. We find that

$$h_m = \frac{1}{F_s} \int_{-F_s/2}^{F_s/2} e^{2\pi jmfT_s} \tilde{H}(f)\, df.$$

As we know that the h_m are even in m, we know that the filter we are designing is not going to be a causal filter—its response to a delta function precedes the arrival of the delta function. We deal with this problem shortly.

Let us consider a simple design example. Suppose that one would like $T_s = 1\,\mathrm{ms}$, and one would like a filter that passes all frequencies up to $100\,\mathrm{Hz}$. That is, $\tilde{H}(f) = 1$ for $|f| \leq 100$, and it is zero otherwise. We find that for $m \neq 0$,

$$\begin{aligned}
h_m &= 0.001 \int_{-500}^{500} e^{2\pi jmf/1000} \tilde{H}(f)\, df \\
&= 0.001 \int_{-100}^{100} e^{2\pi jmf/1000}\, df \\
&= \frac{0.001}{2\pi jm/1{,}000} \left(e^{2\pi jm/10} - e^{-2\pi jm/10} \right) \\
&= \frac{\sin(2\pi m/10)}{\pi m}.
\end{aligned}$$

When $m = 0$, a simple calculation shows that $h_0 = 0.2$.

We find that the coefficients are real and even. Additionally, they extend from $m = -\infty$ to ∞. Clearly, they do not correspond to a *finite* impulse response filter. In order to transform our filter into a finite impulse response filter, we set all the filter coefficients for which $|m| > M$ to zero. That is, we consider the approximation to the ideal filter given by

$$G(z) = \sum_{m=-M}^{M} h_m z^{-m}.$$

What effect does "zeroing" all the coefficients that affect samples far from the current one have? What we have just done is to take a set of filter coefficients and multiply them by a sequence that is one for $|m| \leq M$ and that is zero elsewhere. Defining the *window sequence* $w_m = 1$ for $|m| \leq M$, and zero otherwise, we find that the coefficients of the new filter are $h_m w_m$.

It is not hard to see that multiplying the coefficients in the "time" domain is equivalent to convolution in the frequency domain. Consider two Z-transforms, $H(z)$ and $G(z)$, having regions of convergence that include the unit circle. Let us consider the convolution of $H(e^{2\pi jfT_s})$ with $G(e^{2\pi jfT_s})$

$$H(e^{2\pi jfT_s}) * G(e^{2\pi jfT_s}) \equiv \int_{-F_s/2}^{F_s/2} H(e^{2\pi j\phi T_s})G(e^{2\pi j(f-\phi)T_s})\, d\phi.$$

A quick calculation shows that

$$
\begin{aligned}
H(e^{2\pi jfT_s}) * G(e^{2\pi jfT_s}) &= \int_{-F_s/2}^{F_s/2} H(e^{2\pi j\phi T_s})G(e^{2\pi j(f-\phi)T_s})\, d\phi \\
&= \int_{-F_s/2}^{F_s/2} \left(\sum_{n=-\infty}^{\infty} e^{-2\pi jn(f-\phi)T_s} g_n \right) \times \\
&\qquad \left(\sum_{m=-\infty}^{\infty} e^{-2\pi jm\phi T_s} h_m \right) d\phi \\
&= \sum_{n=-\infty}^{\infty} \sum_{m=-\infty}^{\infty} \int_{-F_s/2}^{F_s/2} e^{-2\pi jn(f-\phi)T_s} g_n e^{-2\pi jm\phi T_s} h_m\, d\phi \\
&= \sum_{n=-\infty}^{\infty} e^{-2\pi jnfT_s} g_n \times \\
&\qquad \sum_{m=-\infty}^{\infty} \int_{-F_s/2}^{F_s/2} e^{-2\pi j(m-n)\phi T_s} h_m g_n\, d\phi \\
&= \sum_{n=-\infty}^{\infty} e^{-2\pi jnfT_s} g_n \sum_{m=-\infty}^{\infty} F_s \delta_{m-n} h_m \\
&= F_s \sum_{n=-\infty}^{\infty} e^{-2\pi jnfT_s} g_n h_n.
\end{aligned}
$$

This is F_s times the frequency response of the system whose coefficients are $h_n g_n$.

In order to determine the effect of the windowing procedure, we must find the function $W(e^{2\pi jfT_s})$ associated with the coefficients w_m. By definition (and after some calculation), we find that

$$
\begin{aligned}
W(e^{2\pi jfT_s}) &= \sum_{k=-\infty}^{\infty} w_k e^{-2\pi jfkT_s} \\
&= \sum_{k=-M}^{M} e^{-2\pi jfkT_s} \\
&= e^{2\pi jMT_s} \sum_{k=0}^{2M} e^{-2\pi jfkT_s} \\
&= e^{2\pi jMT_s} \frac{1 - e^{-2\pi jf(2M+1)T_s}}{1 - e^{-2\pi jfT_s}} \\
&= \frac{\sin(\pi(2M+1)fT_s)}{\sin(\pi fT_s)}.
\end{aligned}
$$

This function is very similar to the sinc function. As M becomes larger, the function tends to T_s times a periodic delta function (as seen in Section 1.2).

When one convolves $\tilde{H}(f)$ with $W(e^{2\pi j f T_s})$, one finds that the resulting function is a smeared version of $\tilde{H}(f)$ that has sidelobes that surround the areas in which the filter response is supposed to be one. As in the case of the calculation of the DFT, the sidelobes are larger when $W(e^{2\pi j f T_s})$ itself has large sidelobes—when it does not decay quickly away from $f = 0$. This, it can be shown, happens when w_m has large abrupt changes. (And this is precisely what we saw in Chapter 5 when we considered the effect that a window has on the DFT of a sequence.) Thus, just as in the case of the DFT, it can be advantageous to use a window function that does not change too abruptly—for example, a raised-cosine (Hann) window. When we used windows to help remove spurious high-frequency terms from the DFT, we found that window functions generally widen the peaks one should see in the DFT. In the case of FIR filter design, window functions generally make the transition from passband to stopband less sharp—and that is a problem *caused* by using window functions.

In Figure 28.1, MATLAB® code that compares the frequency response of the filter designed in Section 28.1 with the frequency response of a filter that uses a windowed version of the same coefficients is given. The figure generated by the code is given in Figure 28.2. We find that the filter implemented without windowing the coefficients has a frequency response with rather pronounced ripple. On the other hand, the frequency response after windowing has a wider transition region.

28.2 Symmetric FIR Filters

If we desire to build a filter that selects certain frequency ranges and that has a constant group delay, then the *symmetric FIR filter* is often a good solution. A symmetric FIR filter is most reasonably defined as an FIR filter for which

$$h_{-n} = h_n.$$

We have seen that if the frequency response we are trying to achieve, $\tilde{H}(f)$, is real and even, then h_n will be real and even as well. Let us consider the transfer function of a symmetric FIR filter. We find that

$$H(z) = \sum_{m=-M}^{M} h_m z^{-m}.$$

Substituting $z = e^{2\pi j f T_s}$, we find that the frequency response of the filter is

$$H(e^{2\pi j f T_s}) = \sum_{m=-M}^{M} h_m (e^{2\pi j f T_s})^{-m}$$

```
M = 100;
% 2M+1 gives the number of taps in the FIR filter.
Ts = 0.001;
% Ts is the sample time.
m = [-M:M]
h = (2 / 10) * sinc(2 * m / 10);
% h contains the filter coefficients.
denom = [zeros([1 M]) 1 zeros([1 M])];
H = tf(h, denom, Ts)
% The filter thus defined is a non-causal symmeteric filter.
figure(1)
bodemag(H,':k',{20,2000})
hwin = h .* hann(2*M+1)';
% Here we window the filter coefficients.
Hwin = tf(hwin, denom, Ts)
% This defines the new filter.
hold on
bodemag(Hwin,'k',{20,2000})
hold off
legend('unwindowed','windowed')
print -djpeg win_com.jpg
% This command generates and saves a JPEG version of the plot.
```

Fig. 28.1. The MATLAB code used to compare an FIR filter whose coefficient values were not windowed with one whose coefficient values were windowed

$$= \sum_{m=-M}^{M} h_m e^{-2\pi jmfT_s}$$

$$= h_0 + h_1(e^{-2\pi jfT_s} + e^{2\pi jfT_s}) + \cdots + h_M(e^{-2\pi jMfT_s} + e^{2\pi jMfT_s})$$

$$= h_0 + h_1 2\cos(2\pi fT_S) + \cdots + h_M 2\cos(2\pi MfT_s).$$

As long as the h_n are real, the frequency response is a real number, and its phase is constant (in intervals). The problem is that this filter is not causal.

The solution is to add a delay to the filter. Let us *redefine* a symmetric FIR filter as a causal $N + 1$-tap (coefficient) filter for which $h_{N-i} = h_i$. If we delay the output of the filter by M samples, then the filter's transfer function is

$$K(z) = \sum_{m=0}^{2M} h_{m-M} z^{-m}.$$

The filter's frequency response is

$$K(e^{2\pi jfT_s}) = \sum_{m=0}^{2M} h_{m-M}(e^{2\pi jfT_s})^{-m}$$

$$= e^{-2M\pi jfT_s} H(e^{2\pi jfT_s})$$

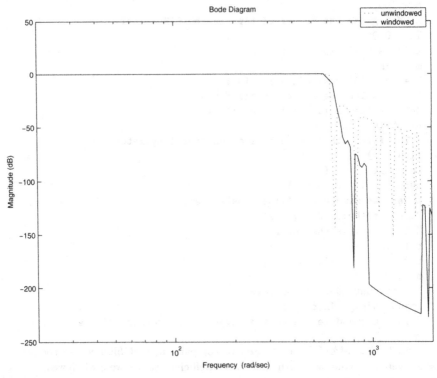

Fig. 28.2. A comparison of the frequency responses of the FIR filter implemented without windowing and the FIR filter implemented with windowing. Note that the windowed coefficients lead to a filter whose frequency response has less ripple away from the transition region, but the unwindowed coefficients lead to a filter with a sharper transition from passband to stopband.

$$= e^{-2M\pi jfT_s}(\text{real number}).$$

Assuming that the real number is positive, which will generally be the case in the regions that are of interest to us, we find that

$$\angle H(e^{2\pi jfT_s}) = -2M\pi fT_s.$$

We find that the group delay of such a filter is

$$\text{group delay} = MT_s.$$

The group delay is *constant* and is exactly M sampling periods long. The fact that the group delay is a constant across all frequencies is the reason we use symmetric FIR filters.

Returning to our 100 Hz filter, we find that to implement it as a causal filter, we must define the filter's transfer function as

$$H(z) = \sum_{m=0}^{2M} h_{m-M} z^{-m}.$$

This filter is causal and, other than a linearly varying phase, has the desired frequency response.

28.3 A Comparison of FIR and IIR Filters

Were it not for the fact that both FIR and IIR filters have their places, one or the other type of discrete-time filter would have been forgotten long ago. FIR filters have two major advantages

1. Symmetric FIR filters have linear phase—and hence the group delay is the same at all frequencies.
2. FIR filters are always stable.

IIR filters, though their phase response is not ideal and they can be unstable have one major advantage: They generally require many fewer taps in their implementation. When using a resource-poor microprocessor, one may find that IIR filters are appropriate. For an interesting discussion that considers many of these points, the reader may wish to consult [16].

28.4 The Experiment

Let us use MATLAB® to start experimenting with FIR filter design. Use the MATLAB command **fdatool** (and the GUI it opens for you) to design a bandpass filter with 99 taps whose passband extends from 25 to 35 Hz. Make the sampling frequency of the filter 200 samples s^{-1}.

When you set up **fdatool**, make sure to select a bandpass filter, make sure that the filter is an FIR filter, and make sure that the design method you choose is "Window." Use several different windows in your design, and print out the frequency response you achieve with each window. Also, analyze the impulse response of the filters. Explain why you see the types of impulse responses you obtain.

Finally, design a nine- or ten-tap bandpass filter that samples at a rate of 1,350 samples s^{-1} and that passes all frequencies in the range extending from 135 to 270 Hz. Window the filter coefficients to keep the frequency response of the filter near zero in the regions where the frequency response is supposed to be zero. Implement the filter on the ADuC841 using a C program, and examine the properties of the filter that has been implemented.

28.5 Exercises

1. Explain what "least squares" FIR filter design is. Feel free to use the library, the web, or any of the various MATLAB "help" features to acquire the information necessary to answer this question.

2. a) Design a 201-tap symmetric high-pass FIR filter for which
   - $T_s = 500\,\mu s$, and
   - the cut-off frequency is 400 Hz.

   b) Have MATLAB calculate and plot the Bode plots corresponding to this filter.

   c) Now, multiply the coefficients by a Hanning window of the appropriate size, and have MATLAB plot the Bode plots of the new filter. How do the new plots compare to those of the unwindowed filter?

3. Please show that for any real even function, the coefficients, b_k, are real and satisfy $b_k = b_{-k}$—they are symmetric in k. You may wish to proceed by showing that for any real function, $b_{-k} = \bar{b}_k$. Then, you may wish to show that for any even function, $b_k = b_{-k}$.

Implementing a Hilbert Filter

Summary. Though many of the filters that we use are frequency-selective—they pick out frequencies that are for one reason or another important to us—not all filters are of this type. Here, we consider the Hilbert filter—an all-pass filter.

Keywords. Hilbert filter, all-pass filter, Fourier series.

29.1 An Introduction to the Hilbert Filter

The frequency response desired of an analog Hilbert filter, $H(f)$, is

$$H(f) = \begin{cases} j, & f < 0 \\ 0, & f = 0 \\ -j, & f > 0 \end{cases}.$$

When implementing the filter digitally, the desired frequency response is

$$\tilde{H}(\Omega) = \begin{cases} j, & -\pi < \Omega < 0 \\ 0, & \Omega = 0 \\ -j, & 0 < \Omega < \pi \end{cases}$$

where $\Omega = 2\pi f T_s$. As this is a discrete-time filter, we know that if $\{h_k\}$ are the filter coefficients, then the filter's frequency response is

$$\text{frequency response} = \sum_{k=-\infty}^{\infty} h_k e^{-j\Omega k}.$$

We choose $\{h_k\}$ such that the frequency response of the filter equals the desired frequency response:

$$\sum_{k=-\infty}^{\infty} h_k e^{-j\Omega k} = \tilde{H}(\Omega).$$

As expected, we find that the h_k are (essentially) the Fourier coefficients of $\tilde{H}(\Omega)$.

Clearly,

$$
\begin{aligned}
h_k &= \frac{1}{2\pi} \int_{-\pi}^{\pi} H(\Omega)e^{j\Omega k}\,d\Omega \\
&= \frac{1}{\pi} \int_{0}^{\pi} \sin(\Omega k)\,d\Omega \\
&= \frac{1}{\pi}\frac{\cos(0) - \cos(\pi k)}{k} \\
&= \begin{cases} \frac{2}{\pi k} & k \text{ odd} \\ 0 & k \text{ even} \end{cases} .
\end{aligned}
$$

Thus, the transfer function of the filter is

$$
H(z) = \sum_{k=-\infty}^{\infty} h_k z^{-k} = \frac{2}{\pi} \sum_{k=-\infty}^{\infty} \frac{1}{2k+1} z^{-(2k+1)}. \tag{29.1}
$$

29.2 Problems and Solutions

As seen in Chapter 28, there are two problems associated with implementing this filter

1. the filter is not causal, and
2. the impulse response is infinite.

We take care of these problems by truncating the response and delaying it. We approximate the Hilbert filter by considering a filter whose transfer function is

$$
H_N(z) = \frac{2}{\pi} z^{-(2N-1)} \sum_{k=-N}^{N-1} \frac{1}{2k+1} z^{-(2k+1)}, \qquad N \geq 1.
$$

29.3 The Experiment

Implement the filter whose transfer function is $H_3(z)$ on the ADuC841. Use a sampling rate of 2,700 samples s^{-1}. In addition to outputting the output of the filter via DAC0, use DAC1 to output the input to the filter, but delay the input by five samples. Compare the output of the two DACs to see the $-90°$ phase shift introduced by the filter. Print out several oscilloscope traces, and include them with your laboratory report.

29.4 Exercises

1. Use MATLAB to show that when $N = 5$, the filter in Section 29.2 provides us with the phase response that we desired (except for a linear term added by the "block" that delays the filter's output by $2 \times 5 - 1$ samples).
2. Discuss the Gibbs phenomenon, and show why it implies that approximations of the type we are making will never tend uniformly to the magnitude response we want. Show that there will always be points of the magnitude response that differ from the desired value by several percent.
3. Is the filter given by (29.1) stable? Explain!

30

The Goertzel Algorithm

Summary. We have seen that the FFT allows one to calculate the DFT of an N-term sequence in $O(N \ln(N))$ steps. As calculating a single element of the DFT requires $O(N)$ steps, it is clear that when one does not need too many elements of the DFT, one is best off calculating individual elements, and not the entire sequence. In this chapter, we present a simple algorithm—the Goertzel algorithm—for calculating individual elements of the DFT.

Keywords. Goertzel algorithm, DFT, second-order filter.

30.1 Introduction

Consider the definition of the DFT

$$Y_m = \mathrm{DFT}(\{y_k\})(m) \equiv \sum_{k=0}^{N-1} e^{-2\pi jmk/N} y_k.$$

The calculation of any given coefficient, Y_m, takes $O(N)$ steps. Thus, if one only needs a few coefficients (fewer than $O(\ln(N))$ coefficients), then it is best to calculate the coefficients and not bother with the "more efficient" FFT algorithm (which calculates *all* of the Fourier coefficients). The Goertzel algorithm[1] is a simple way of calculating an individual Fourier coefficient. It turns calculating a Fourier coefficient into implementing a second-order filter and using that filter for a fixed number of steps. The Goertzel algorithm is somewhat more efficient than a "brute force" implementation of the DFT.

30.2 First-order Filters

Consider the solution of the equation

---

[1] Named after its discoverer, Gerald Goertzel. The algorithm was published in 1958 [11].

$$r_n = \alpha r_{n-1} + x_n. \tag{30.1}$$

This corresponds to calculating the response of the filter whose transfer function is

$$\frac{R(z)}{X(z)} = \frac{z}{z - \alpha}.$$

Making use of the variation of parameters idea [17], we guess that the solution of (30.1) is of the form

$$r_n = \alpha^n z_n.$$

We find that we must produce a z_n for which

$$
\begin{aligned}
r_n &= \alpha^n z_n \\
&= \alpha r_{n-1} + x_n \\
&= \alpha(\alpha^{n-1} z_{n-1}) + x_n \\
&= \alpha(\alpha^{n-1}(z_n + (z_{n-1} - z_n))) + x_n \\
&= \alpha^n z_n + \alpha^n(z_{n-1} - z_n) + x_n.
\end{aligned}
$$

In order for equality to hold, we find that

$$z_n = z_{n-1} + \alpha^{-n} x_n. \tag{30.2}$$

Assuming that $r_n = x_n = 0$ for $n < 0$, (30.2) implies that

$$z_n = \sum_{k=0}^{n} \alpha^{-k} x_k.$$

Finally, we find that

$$r_n = \alpha^n z_n = \sum_{k=0}^{n} \alpha^{n-k} x_k. \tag{30.3}$$

(For another way of arriving at this result, see Exercise 3.)

30.3 The DFT as the Output of a Filter

Let us consider the definition of the DFT again. We find that

$$
\begin{aligned}
Y_m &= \sum_{k=0}^{N-1} e^{-2\pi j m k/N} y_k \\
&= \sum_{k=0}^{N-1} (e^{2\pi j m/N})^N (e^{-2\pi j m/N})^k y_k.
\end{aligned}
$$

$$= \sum_{k=0}^{N-1} (e^{2\pi jm/N})^{N-k} y_k$$

$$= e^{2\pi jm/N} \sum_{k=0}^{N-1} (e^{2\pi jm/N})^{N-1-k} y_k.$$

We find that Y_m is the output of a first-order filter at "time" $N - 1$.

Let us examine the relevant filter. We find that the filter's transfer function is

$$T(z) = \frac{ze^{2\pi jm/N}}{z - e^{2\pi jm/N}}.$$

This transfer function can be rewritten as

$$T(z) = \frac{z(ze^{2\pi jm/N} - 1)}{z^2 - 2z\cos(2\pi m/N) + 1}.$$

This transfer function can be considered the product of two transfer functions, $T_1(z)T_2(z)$, where

$$T_1(z) = \frac{1}{1 - 2z^{-1}\cos(2\pi m/N) + z^{-2}}, \text{ and } T_2(z) = e^{2\pi jm/N} - z^{-1}.$$

We find that the first transfer function has real coefficients, and the second corresponds to a finite impulse response filter. As the output of the second filter is needed only when $k = N - 1$, the FIR filter need not be implemented except at that last step.

We can calculate Y_m by implementing the filter

$$r_n = 2r_{n-1}\cos(2\pi m/N) - r_{n-2}$$

and continuing the calculation until $n = N - 1$. At that point, we calculate $Y_m = e^{2\pi jm/N} r_{N-1} - r_{N-2}$. This algorithm is known as the Goertzel algorithm.

30.4 Comparing the Two Methods

To what extent is the Goertzel algorithm more efficient than the direct calculation of the value of Y_m,

$$Y_m = \sum_{k=0}^{N-1} e^{-2\pi jmk/N} y_k?$$

If one performs a brute force calculation of Y_m, one must multiply $e^{-2\pi jmk/N}$ by y_k for N values of k. As the complex exponential is essentially a pair of real numbers, this requires $2N$ real multiplications. Additionally, the sum requires that $2(N-1)$ real sums be calculated. When using the Goertzel algorithm, the recurrence relation requires two real additions and one real multiplication at each step. The FIR filter that is used in the final stage requires two real multiplications and one addition. In sum, the Goertzel algorithm requires $N+2$ real multiplications and $2N+1$ real additions. We find that the Goertzel algorithm is somewhat more efficient than the brute force calculation.

30.5 The Experiment

Implement the Goertzel algorithm using Simulink®. Calculate one of the elements of the FFT of a sequence that is 16 elements long.

1. Implement the filter $T_1(z)$. As its coefficients are all real, this is not a problem.
2. Implement filter $T_2(z)$ as two separate filters. One should produce the real part of the FFT, and the second should produce the imaginary part.
3. Let the input to the filters come from the workspace, and let the output of each of the filters go to the workspace. (Look at the Simulink sources and sinks to find the appropriate blocks.)
4. Try out the final set of filters with a variety of inputs. Note the size of the output of the filter $T_1(z)$.
5. In your report, explain why implementing the Goertzel algorithm using a microprocessor might be challenging.

30.6 Exercises

1. Discuss the stability of filter $T_1(z)$.
2. Why are we able to use $T_1(z)$ even though it is a "problematic" filter? Explain your answer clearly.
3. Show that (30.3) describes the relation between x_n and r_n by using the fact that the operations being described are causal, by assuming that $x_n = 0$ when $n < 0$, and by making use of the Z-transform.

References

1. Aziz, P.M., Sorensen, H.V., and van der Spiegel, J., "An Overview of Sigma–Delta Converters," *IEEE Signal Processing Magazine*, Vol. 13, No. 1, 1996.
2. Bennett, S., "Control Systems: The Classical Years 1935-1955," http://www.mech.gla.ac.uk/Research/Control/Seminars/bennett.html, accessed June 7, 2007.
3. Churchill, R. V., Brown, J. W., and Verhey, R. F., *Complex Variables and Applications*, 3rd edn., McGraw-Hill, New York, 1976.
4. Constantinides, A., "Spectral Transformations for Digital Filters," *Proc. IEEE*, Vol. 117, No. 8, 1970.
5. Cooley, J.W., and Tukey, J.W., "An Algorithm for the Machine Calculation of Complex Fourier Series," *Math. Comput.*, Vol. 19, 1965.
6. Engelberg, S., *A Mathematical Introduction to Control Theory*, Series in Electrical and Computer Engineering, Vol. 2, Imperial College Press, London, 2005.
7. Engelberg, S., *Random Signals and Noise: A Mathematical Introduction*, CRC Press, Boca Raton, FL, 2006.
8. Engelberg, S., "Implementing a Sigma–Delta DAC in Fixed Point Arithmetic," *IEEE Signal Processing Magazine*, Vol. 23, No. 6, 2006.
9. Engelberg, S., *A Microprocessor Laboratory Using the ADuC841*, Jerusalem College of Technology, Jerusalem, Israel, 2007, http://www.springer.com/978-1-84800-118-3.
10. Frantz, G. and Simar, R., "DSP: Of Processors and Processing," *ACM Queue*, Vol. 2, No. 1, 2004, http://acmqueue.com/modules.php?name=Content&pa=showpage&pid=125, accessed November 12, 2007.
11. Goertzel, G., "An Algorithm for the Evaluation of Finite Trigonometric Series," *American Mathematical Monthly*, Vol. 65, No. 1, 1958.
12. Jung, W., Editor, *Op Amp Applications Handbook*, Newnes, Burlington, MA, 2005, http://www.analog.com/library/analogDialogue/archives/39-05/op_amp_applications_handbook.html, accessed June 11, 2006.
13. Keil Elektronik GmbH., *Cx51 Compiler: Optimizing C Compiler and Library Reference for Classic and Extended 8051 Microcontrollers*, 1988-2000.
14. Oppenheim, A.V., and Schafer, R.W., *Digital Signal Processing*, Prentice-Hall, Englewood Cliffs, NJ, 1975.

15. Oppenheim, A.V., Willsky, A.S., and Young, I.T., *Signals and Systems*, Prentice-Hall, Englewood Cliffs, NJ, 1983.

16. Rader, C.M., "DSP History – The Rise and Fall of Recursive Digital Filters," *IEEE Signal Processing Magazine*, Vol. 23, No. 6, 2006.

17. Thomas, G.B., *Calculus and Analytic Geometry*, 4th edn., Addison Wesley, Reading, MA, 1968.

18. Wikipedia, the free encyclopedia, http://en.wikipedia.org, accessed November 22, 2007.

19. Zemanian, A.H., *Distribution Theory and Transform Analysis*, Dover, New York, 1987.

Index